Women in Engineering and Science

Series Editor

Jill S. Tietjen, Greenwood Village, USA

The Springer Women in Engineering and Science series highlights women's accomplishments in these critical fields. The foundational volume in the series provides a broad overview of women's multi-faceted contributions to engineering over the last century. Each subsequent volume is dedicated to illuminating women's research and achievements in key, targeted areas of contemporary engineering and science endeavors.The goal for the series is to raise awareness of the pivotal work women are undertaking in areas of keen importance to our global community.

Kaitlyn J. Bunker
Editor

Women in Electrical Engineering

Springer

Editor
Kaitlyn J. Bunker
Boulder, CO, USA

ISSN 2509-6427 ISSN 2509-6435 (electronic)
Women in Engineering and Science
ISBN 978-3-032-00132-0 ISBN 978-3-032-00133-7 (eBook)
https://doi.org/10.1007/978-3-032-00133-7

This Springer imprint is published by the registered company Springer Nature Switzerland AG
The registered company address is: Gewerbestrasse 11, 6330 Cham, Switzerland

For Ellison and Emrin—Kaitlyn J. Bunker

Foreword

The technical communities of electrical engineering and its interconnected domains of power and energy, waves and signals, circuits and systems, communications and computing have driven much of modern society's transforming technological innovations. Despite its dominant influence in effects, the harnessed motions of electrons are rarely seen directly. Engineering enthusiasm for these applications has had a less direct path for many, often growing from an initial inspiration rooted in math, science, or an engineering field that is easier to see and touch. Electrical engineering, in some ways perhaps due to this, historically and presently faces a significant lack of female representation. *Women in Electrical Engineering*, edited by Kaitlyn Bunker, pulls to the front both the visible-by-impact field and the important contributions made by women to historic and emerging technologies, as told by several knowledgeable authors who have joined in this effort to share firsthand their expertise and offer some inspiring achievements.

I have the honor of serving as the 2025 IEEE President and CEO, a global organization of 500,000 members whose roots trace back 141 years ago to the American Institute of Electrical Engineers. Then, it was Western Union's Norvin Green who served as the first president and Thomas Alva Edison as the first vice president. While I am not the first female IEEE President but the fifth, it was a 109-year journey to Martha Sloan achieving that distinction in 1993.

Across many professions and in many fields of advanced studies, women have at long last become better represented, even coming to make up the majority of the next generation. This is true in medicine, life and health sciences, social sciences, liberal arts, accounting, and law. In engineering, it remains less so, and especially in a few fields that face the greatest challenges by far: electrical and computer, mechanical, and aerospace engineering.

One of my more joyful roles as IEEE President is to celebrate and promote recognition of the history of technological innovation and the impacts on society of the innovators and their pioneering advances for and from electrical engineering fields that have and will continue to change the future. The first chapter, "Pioneering Women in Electrical Engineering," develops the field itself by showcasing the impact made to the field by prominent women, while a later chapter focuses on the more contemporary career path of Anne Lucietto and the discoveries made across industries.

I love to learn, and some chapters offer technical topics to expand your knowledge. Radomes was a new word to me and is a theme for one chapter, which illustrated to me the difference between learning your electromagnetic fields from engineers versus from physicists, as I did in my student days. Electrical engineering does rely on what is not directly seen and moved, and radomes as an engineering challenge is to encase large antennas in ways that are invisible to the signals they are intended to receive. A chapter also showcases antennas and the breadth of different engineering applications and different scales of technologies that rely upon antennas as they send or receive electromagnetic waves for communication, detection, and even power.

Chapters on personal resilience, career versatility, powering change, and breaking barriers provide wisdom and lessons told with personal stories that draw from and relate to, respectively, each of the authors' own stories of technical knowledge and challenges over the long term. These perspectives are offered from experts in fields such as microgrids, aerospace systems, renewable energy, and careers in the utility industry.

Enjoy!

2025 IEEE President and CEO
University of San Diego
San Diego, CA, USA

Kathleen A. Kramer

Contents

About the Editor

Kaitlyn J. Bunker, Ph.D., P.E. is a Director with RMI's Global South program, leading the Climate Finance Access Network (CFAN). In this role, she leads the team at RMI that serves as network coordinator for CFAN, which aims to unlock access to climate finance at scale in developing countries. Kaitlyn previously worked with RMI's Islands Energy Program, leading a diverse team partnering with islands in the Caribbean to support and accelerate their clean energy transitions.

Before joining RMI, Kaitlyn completed a Ph.D. in electrical engineering from Michigan Technological University in Houghton, MI. Her dissertation research focused on microgrids and optimizing control strategies for distributed renewable resources. She stayed at Michigan Tech for graduate school following her undergraduate studies, which were also in electrical engineering.

With a mechanical engineer father and biochemist mother, Kaitlyn knew from a young age that she wanted to pursue a science, technology, engineering, and math (STEM) career. She received advice that electrical engineers use the most math out of all of the engineering disciplines, and as a math lover, she decided to pursue this option.

As a college student and later as a professional, Kaitlyn has been a longtime member of the Society of Women Engineers (SWE). Participating in SWE has allowed her to gain new leadership skills while contributing to an organization whose mission she is proud of. As a Ph.D. student, Kaitlyn served as the Collegiate Director on SWE's Board of Directors, which opened up a new understanding of the importance of strategic thinking and collaboration. Kaitlyn received the SWE Distinguished New Engineer award in 2018 and served again on SWE's Board of Directors as Secretary during 2023–2025 and as a Director during 2025–2027.

Kaitlyn lives in Colorado with her husband and two daughters. They are slowly renovating their home to be as energy efficient as possible, with a goal to eventually be net-zero energy.

Pioneering Women in Electrical Engineering

Jill S. Tietjen

1 Introduction

Women in Europe and the USA in the 1700s and 1800s laid the foundation for later generations of women to pursue careers in all of the science, technology, engineering, and mathematics (STEM) fields including electrical engineering. They include science popularizers Jane Haldimand Marcet and Mary Fairfax Somerville; science writer and educator Almira Hart Lincoln Phelps; and computer software pioneer Ada Byron Lovelace. From the time Bertha Lamme graduated from The Ohio State University in 1893 through today, women have made their mark in the field of electrical engineering. Lamme worked on the turbo generators for the hydroelectric plant at Niagara Falls and the motors that operated the power plant of the Manhattan Elevated Railroad. Although Edith Clarke was the first woman to receive a master's degree in electrical engineering from MIT, she had difficulty finding a job due to her gender—there were no job openings for women engineers! Later, General Electric did hire her and she made seminal contributions to the analysis of transmission lines for electric utilities. Many women in electrical engineering have followed in the footsteps of these early pioneers. A few of their accomplishments include opening up research opportunities for women in the Antarctic, helping to create the electrical installations at the Hoover Dam, developing very large scale integration (VLSI) design methodology for chip making, and inventing what today is known as RealD 3D for 3D films. These pioneers have also been fierce advocates for women, helping women succeed in the STEM fields.

J. S. Tietjen (✉)
Technically Speaking, Inc., Greenwood Village, CO, USA

K. J. Bunker (ed.), *Women in Electrical Engineering*, Women in Engineering and Science, https://doi.org/10.1007/978-3-032-00133-7_1

2 Jane Haldimand Marcet (1769–1858)

Briton Jane Marcet (Fig. 1) worked to popularize science—writing books and papers that explained science to general readers—specifically intended for women and young people. She is particularly remembered for the impact her *Conversations in Chemistry* had on influencing future scientist Michael Faraday (for whom the unit of electrical capacitance is named).[1] Jane was encouraged to begin a writing career by her husband, physician Dr. Alexander Marcet, whose passion for chemistry exceeded his interest in practicing as a physician. *Conversations in Chemistry* (1806) was very popular and went through numerous editions, including 15 American editions titled *Mrs. Bryan's Conversations*. Jane Marcet believed that the information presented in a conversational format was more readily comprehended by the audience, as she was better able to understand chemistry after conversing with a friend. Her other books included *Conversations on Botany*, *Conversations on Natural Philosophy, Conversations on Political Economy*, and *Conversations on Vegetable Physiology* [1–5].

Fig. 1 Jane Haldimand Marcet. (Courtesy of Wikipedia)

[1] Pioneering physicist Michael Faraday produced the first electrical generators and motors. The farad, the unit of electrical capacitance, was named for him. He introduced the terms "electrolyte" and "ions" into our scientific vocabulary.

3 Mary Fairfax Somerville (1780–1872)

One of the first honorary women members of the Royal Astronomical Society (Great Britain), Briton Mary Somerville (Fig. 2), like Jane Haldimand Marcet, helped to popularize science particularly for women. After spending a year at a boarding school when she was 10 years old, Mary developed a thirst for reading and arithmetic. She taught herself Latin, and then algebra, after seeing strange symbols in a ladies' fashion magazine. When her parents found out about her interest in mathematics, her father forbade her study due to worries that mental activity would harm her female body. After her first husband died and left her with a modest inheritance, she openly educated herself in trigonometry and astronomy. Most of her friends and family did not support her educational efforts.

Mary married her first cousin, Dr. William Somerville, and found in him someone to support her pursuits of educational and intellectual matters. In fact, William encouraged Mary to expand her studies beyond mathematics and astronomy to Greek, botany, and mineralogy. In 1834, she published *On the Connexion of the Physical Sciences* which presented a comprehensive picture of the latest research in the physical sciences. Her 1831 book, *Mechanism of the Heavens*, contributed to the modernization of English mathematics. Mary was occasionally criticized for her "unwomanly" pursuit of science, nevertheless, she was referred to, both in England and abroad, as "the premier scientific lady of the ages" [6].

Fig. 2 Mary Fairfax Somerville. (Courtesy Library of Congress)

4 Almira Hart Lincoln Phelps (1793–1884)

Almira Hart Lincoln (Fig. 3) became interested in science after the death of her husband in 1823 and her subsequent return to teaching on the staff of her sister's (Emma Hart Willard) school, the Troy Female Seminary. She was encouraged by Rensselaer professor Amos Eaton in her scientific pursuits. In 1829, she published her first science textbook, *Familiar Lectures on Botany*. After she married John Phelps in 1831, she continued to write and revise her textbooks in addition to taking care of her family responsibilities. Other volumes included *Chemistry for Beginners* (1834) and *Familiar Lectures on Natural Philosophy* (1837).

She served as the principal for several female seminaries and emphasized science in the curriculum. Her teaching innovations included experimental methods in chemistry and botany. In 1859, she was the third woman elected as a member of the American Association for the Advancement of Science. Although she supported educational equality for women, she opposed suffrage and was active in the Woman's Anti-Suffrage Association. Upon her death, her herbarium of plants, collected throughout her lifetime, was presented to the Maryland Academy of Sciences, of which she was the first woman member [2, 3, 7, 8].

Fig. 3 Almira Hart Lincoln Phelps. (Courtesy of Wikipedia)

5 Ada Byron Lovelace (1815–1852)

The daughter of the English poet Lord George Byron, Ada Lovelace now has a computer language named (Ada) after her. A somewhat sickly child, Lovelace was tutored at home and was competent in mathematics, astronomy, Latin, and music by the age of 14. Totally enthralled by Charles Babbage's Difference Engine (an early computer concept), at 17 years old, Lovelace began studying differential equations. As proposed, his second machine, the analytical engine, could add, subtract, multiply, and divide directly and it would be programmed using punched cards, the same logical structure used by the first large-scale electronic digital computers in the twentieth century.

In 1842, the Italian engineer, L.F. Menabrea published a theoretical and practical description of Babbage's analytical engine. Lovelace translated this document adding "notes" in the translation. Her notes constitute about three times the length of the original document and, as explained by Babbage, the two documents together show "That the whole of the development and operations of analysis are now capable of being executed by machinery." These notes include a recognition that the engine could be told what analysis to perform and how to perform it—the basis of computer software. Her notes (Fig. 4) were published in 1843 in *Taylor's Scientific Memoirs* under her initials, because although she wanted credit for her work, it was considered undignified for aristocratic women to publish under their own names.

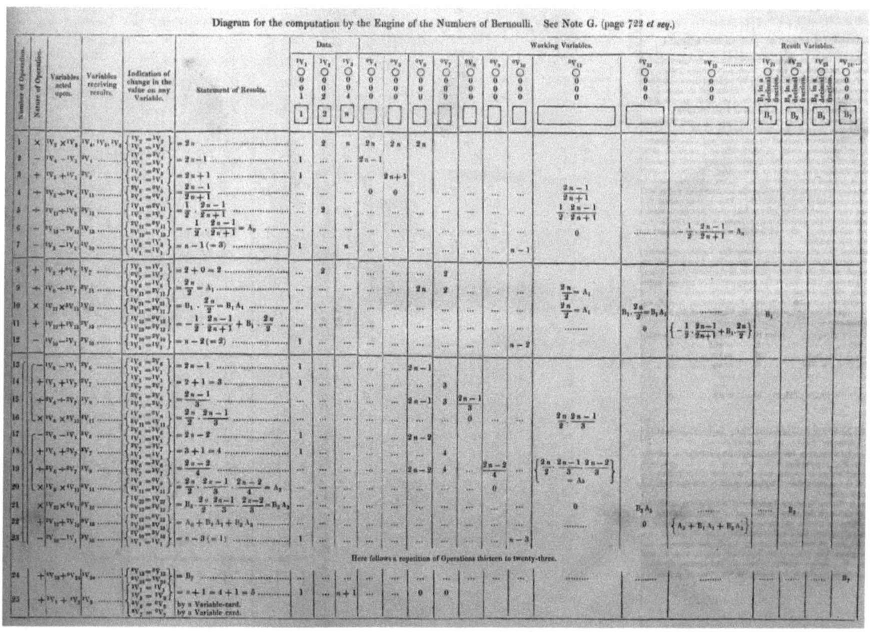
Diagram for the computation by the Engine of the Numbers of Bernoulli. See Note G. (page 722 et seq.)

Number of Operation. Nature of Operation. Variables acted upon. Variables receiving results. Indication of change in the value on any Variable. Statement of Results. Data. Working Variables. Result Variables.

Here follows a repetition of Operations thirteen to twenty-three.

Fig. 4 Ada Byron Lovelace—Note G. (Courtesy of Wikipedia)

Ada Lovelace is considered to be the first person to describe computer programming [9, 10].

6 Bertha Lamme (Feicht) (1869–1943)

The first woman to graduate with a degree in engineering other than civil engineering, Bertha Lamme (Fig. 5) graduated in mechanical engineering (with an electrical engineering emphasis) from The Ohio State University in 1893, the second woman to receive an engineering degree at all and the only woman in her class. Lamme was considered an expert in motor design and had joined her brother (who had also graduated in mechanical engineering from The Ohio State University) at the Westinghouse Electric and Manufacturing Company in Pittsburgh, Pennsylvania. She was a member of her brother's team at Westinghouse and their projects together included the first turbo generators for the hydroelectric plant at Niagara Falls and the motors that operated the power plant of the Manhattan Elevated Railroad. An unpublished document from Westinghouse files refers to Lamme as a "designing engineer" and "one of the few women who have made a notable success of this work."

Fig. 5 Bertha Lamme (George Westinghouse Museum Collection, Detre Library & Archives, Senator John Heinz History Center)

Although she was part of a team and restricted from the shop floor or the field due to her gender, the *Pittsburgh Dispatch* in 1907 reported:

> … even in that hothouse of gifted electricians and inventors. She is accounted a master of the slide rule and can untangle the most intricate problems in ohms and amperes as easily and quickly as any man expert in the shop.

Earlier, in December 1899, the *Woman's Journal*—based on a report in the *New York Sun* from the previous month—called Lamme "the particular star among American women electricians," and noted that she "designs machinery, makes calculations, and does exactly the work of a male electrical engineer" [11].

Both her brother and her husband attained fame for their accomplishments and it is not clear how much she helped either one. Her brother lived with Lamme and her husband throughout the remainder of his life. Lamme remained at Westinghouse until 1905 when she married her supervisor (and was required to leave employment). The Ohio State University named the Lamme Power Systems Laboratory in honor of Bertha and her brother Benjamin. Lamme's slide rule and a drawing she made of a drill bit are in the permanent "Pittsburgh: A Tradition of Innovation" exhibit at the Senator John Heinz History Center. The Center opines "A woman in a man's world. Never before had a female sat at the drafting table with men to compute, calculate and design the tools, motors and machinery that powered the new era of electricity." The Society of Women Engineers (SWE) annually awards a Bertha Lamme Memorial Scholarship, established in 1973, in conjunction with the Westinghouse Educational Foundation. Her daughter became a physicist [3, 12–14].

7 Edith Clarke (1883–1959)

A woman engineer with many firsts to her name, Edith Clarke (Fig. 6) grew up in Maryland without any intentions of even going to college. After graduating from Vassar with an A.B. in mathematics and astronomy in 1908 (Phi Beta Kappa), Clarke taught math and science for 3 years in San Francisco and West Virginia. But, teaching was not holding her interest and she decided to pursue becoming an engineer instead. She enrolled as a civil engineering undergraduate student at the University of Wisconsin and remained there for a year. Then, she went to work for American Telephone & Telegraph Company (AT&T) as a computing assistant. She intended to return to the University of Wisconsin to complete her engineering studies but found the work so interesting at AT&T that she stayed for 6 years.

During World War I, she supervised the women at AT&T who did computations for research engineers in the Transmission Department. She simultaneously studied radio at Hunter College and electrical engineering at Columbia University at night. Eventually, she enrolled at MIT and received her master's degree in electrical engineering in 1919; the first woman awarded that degree from MIT. Upon graduation, she wanted to work for either General Electric (GE) or Westinghouse. But even with her stellar credentials, no one would hire her as an engineer because of her

Fig. 6 Edith Clarke. (Photograph courtesy of Society of Women Engineers Records, Walter P. Reuther Library, Wayne State University)

gender—they had no openings for a woman engineer! In 1920, after a long job search, GE offered Clarke a computing job, directing women computers who were calculating the mechanical stresses in turbines for the turbine engineering department at GE.

But, Clarke wanted to be an electrical engineer! Since that was not the job she was offered and since she wanted to travel the world, she left GE in 1921 to teach physics at the Constantinople Women's College (now Istanbul American College) in Turkey. A year later GE did offer her a job as an electrical engineer in the central station engineering department. When she accepted this job, she became the first professionally employed female electrical engineer in the USA.

Clarke's area of specialty was electric power systems and problems related to their operation. She made innovations in long-distance power transmission and the development of the theory of symmetrical components and circuit analysis. Symmetrical components are a mathematical means by which engineers can study and solve problems of power system losses and performance of electrical equipment. Clarke literally wrote the textbook: *Circuit Analysis of AC Power Systems, Symmetrical and Related Components* (1943) and a second volume in 1950. This textbook, in its two volumes, was used to educate all power system engineers for many years.

She published 18 technical papers during her employment at GE reflecting her status as an authority on the topics of hyperbolic functions, equivalent circuits, and graphical analysis within electric power systems. "Simplified Transmission Line Calculations," which appeared in the *General Electric Review* in May 1926, provided charts for transmission line calculations. She was also involved in the design of hydroelectric dams in the Western USA.

Clarke received a patent in 1925 (1,552,113) for her "graphical calculator"—a method of considering the impacts of capacity and inductance on long electrical transmission lines (Fig. 7). It greatly simplified the calculations that needed to be

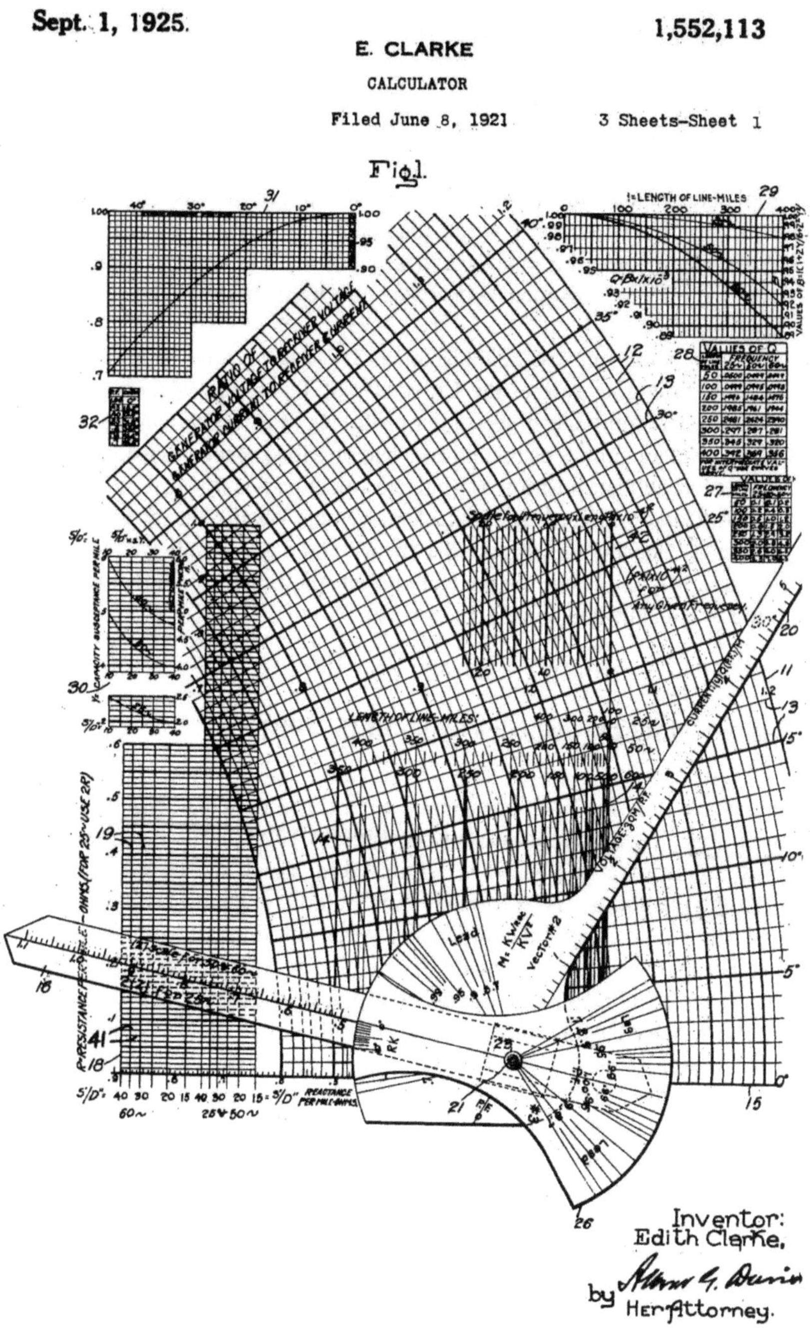

Fig. 7 Edith Clarke's patent number 1,552,113 calculator

done. In 1926, she was the first woman to address what is today the Institute of Electrical and Electronics Engineers (IEEE)—at the time it was the American Institute of Electrical Engineers (AIEE). Her topic was "Steady-State Stability in Transmission Systems." In 1932, Clarke became the first woman to present a paper before the AIEE; her paper, "Three-Phase Multiple-Conductor Circuits," was named the best paper of the year in the northeastern district. This paper examined the use of multiple conductor transmission lines with the aim of increasing the capacity of the power lines. In 1948, Clarke was named one of the first three women fellows of IEEE. She had previously become the first female full voting member of IEEE. Clarke was one of the few women who were licensed professional engineers in New York State.

A year after her retirement from GE in 1945, Clarke became an associate professor of electrical engineering at the University of Texas. In 1947, she rose to full professorship becoming the first woman professor of electrical engineering in the USA. She served in numerous committees and provided special assistance to graduate students through her position as graduate student advisor.

In 1954, Clarke received the SWE Achievement Award "in recognition of her many original contributions to stability theory and circuit analysis." In 2015, she was posthumously inducted into the National Inventors Hall of Fame for her invention of the graphical calculator [3, 11, 14–17].

8 Mabel MacFerran Rockwell (1925–1981)

Credited as possibly the first aeronautical engineer in the USA, Mabel MacFerran Rockwell (Fig. 8) received her BS from MIT in 1925 and a BS from Stanford University in electrical engineering. Before World War II, she served as a technical assistant with the Southern California Edison Company, where she was a pioneer in the application of symmetrical components to transmission relay problems in power systems. Through this work, she made it easier to diagnose system malfunctions and to enhance the reliability of multiple-circuit lines. Rockwell then worked for the Metropolitan Water District in Southern California where she was a member of the team that designed the Colorado River Aqueduct's power system and the only woman to participate in the creation of the electrical installations at the Hoover Dam (previously called the Boulder Dam).

Later, Rockwell joined Lockheed Aircraft Corporation and worked to improve the manufacturing operations of aircraft. Her many innovations included refining the process of spotwelding and developing techniques for maintaining cleaner working surfaces so that the welds completely fused. After the war, Rockwell went to work for Westinghouse where she designed the electrical control system for the Polaris missile launcher. At Conair, she developed the launching and ground controls for the Atlas guided missile systems. In 1958, President Eisenhower named her Woman Engineer of the year. Also in 1958, she received SWE's Achievement Award

Fig. 8 Mabel MacFerran Rockwell. (Photograph courtesy of Society of Women Engineers Records, Walter P. Reuther Library, Wayne State University)

"in recognition of her significant contributions to the field of electrical control systems" [14, 18].

9 Irene Peden (1925–)

The first American woman to live and work in the interior of the Antarctic, Dr. Irene Peden (Fig. 9) pioneered the use of radio waves to analyze deep glacial ice. After earning her electrical engineering degrees from the University of Colorado and Stanford University (the first PhD in electrical engineering awarded to a woman at Stanford), Dr. Peden began her academic career on the electrical engineering faculty at the University of Washington, the first woman to join the entire engineering faculty. She spent her entire career at that university.

The first woman to spend an entire winter at the South Pole, Dr. Peden was almost prevented from taking the flight from Christchurch, New Zealand to the Antarctic in 1970 as the Navy required that another woman accompany her. The woman who had planned to go had not passed her physical; but another woman was found—not a scientist but a librarian who was a mountaineer. On that trip, Peden's team was the first to measure electrical properties of the Antarctic ice. Later her research would expand to include using radio waves to determine ice thickness and to search for structures under the ice. Told that if her experiment did not succeed, women probably wouldn't be allowed on the Antarctic continent for another generation, Peden felt she could not fail. "Well, that's a difficult thing to tell a person doing experimental work, because if it's experimental and it's really research, you don't

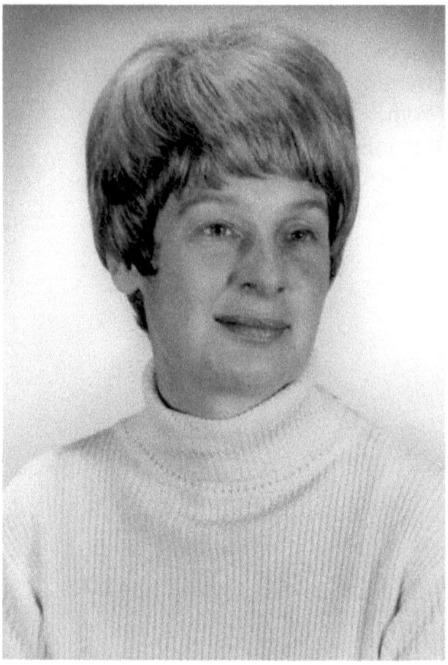

Fig. 9 Irene Peden. (Photograph courtesy of Society of Women Engineers Records, Walter P. Reuther Library, Wayne State University)

know how it's going to turn out until it does. So that was a bit of a risk, but I was quite willing to take it. I thought I knew what I was doing." Her experiment was successful and today Peden Cliffs on the continent is named in her honor.

Her many awards include the "Man of the Year" Award from the IEEE Antennas and Propagation Society, the National Science Foundation's Engineer of the Year, induction into the American Society for Engineering Education Hall of Fame, and the Society of Women Engineer's 1973 Achievement Award "In recognition of her significant contributions in the fields of radio wave propagation research and electrical engineering education." A member of the National Academy of Engineering, Peden is an IEEE Fellow and has received its Third Millenium Medal and its Centennial Medal [19, 20].

10 Harriett Rigas (1934–1989)

Harriett Rigas (Fig. 10), who received her master's and PhD degrees from the University of Kansas, was a strong advocate for women in engineering. Originally from Canada, she pursued an academic career at Washington State University and Michigan State University after employment at the Mayo Clinic and the Lockheed Missile and Space Company. Instrumental in the development of automatic software patches (allowing for upgrades instead of reinstalling the program), Rigas was recognized for her hybrid computer and computer simulation research. She also worked to create codes that computers used to store and access data. At Washington State, she developed the computer engineering curriculum and served as a faculty

Fig. 10 Harriett Rigas (second from right) receiving her Achievement Award from the Society of Women Engineers. (Photograph courtesy of Society of Women Engineers Records, Walter P. Reuther Library, Wayne State University)

member at the Naval Postgraduate School. She chaired engineering departments at both universities.

Rigas was named an IEEE Fellow in 1984 "for contributions to programming of analog/hybrid computers and the development of computer engineering curricula." Her other awards included the 1982 Achievement Award from the Society of Women Engineers "for significant contributions in the field of electrical engineering and computer technology" and the University of Kansas' Distinguished Engineering Service Award. She was very active in the engineering community serving on numerous boards and committees. Posthumously, she has been honored at Washington State through the Harriett B. Rigas Memorial Scholarship Fund and the Harriett B. Rigas Outstanding Woman in Doctoral Studies Award, at Michigan State through the naming of the Harriett B. Rigas Graduate Engineering Program, and by the IEEE through the Harriett B. Rigas Award which recognizes outstanding faculty women who have made a significant contribution to undergraduate education [21–23].

11 Lynn Conway (1938–2024)

Today, what is called the Mead-Conway very large scale integration (VLSI) design methodology was radical when proposed by Lynn Conway (Fig. 11) and Carver Mead. In the 1970s, they worked to prototype and write the textbook on this design methodology which has formed the foundation for global chip making.

Fig. 11 Lynn Conway. (Courtesy of Wikipedia)

Lynn Conway grew up as a curious child with a deep love of books and adventure. She earned bachelor's (1962) and master's (1963) degrees in electrical engineering at Columbia University before joining IBM Research. Her seminal contributions to computer architecture there included multiple out-of-order dynamic instruction scheduling. When Conway decided to transition from male to female, she was fired by IBM. She changed her identity and worked for several companies before joining Xerox PARC. It was at Xerox PARC that she collaborated with Mead on what has become the Mead-Conway VLSI design methodology. After she and Mead developed VLSI and wrote the textbook, she taught the first course on VLSI at MIT as an associate visiting professor in 1978. By 1983, about 120 schools were teaching the methodology.

Conway joined the University of Michigan in 1985 and taught there for many years. She retired from active teaching and research in 1998. Conway began her transgender activism in 1999 after learning that her story was going to be published. In 2020, IBM officially apologized to her for firing her.

Conway's many honors include induction into the National Inventors Hall of Fame and election to the National Academy of Engineering. The recipient of the Society of Women Engineers' Achievement Award in 1990 "For essential contributions to very large-scale integrated (VLSI) circuit and system design methodology, and for rapid propagation of innovations throughout the engineering community," she also received the IBM Lifetime Achievement Award. She was a Fellow of the Computer History Museum "For her work in developing and disseminating new methods of circuit design" [24, 25].

Fig. 12 Martha Sloan. (Photograph courtesy of Society of Women Engineers Records, Walter P. Reuther Library, Wayne State University)

12 Martha Sloan (1939–)

The first female President of the IEEE, Martha Sloan (Fig. 12) was also the first female member of the electrical engineering faculty at Michigan Technological University. Her research interests included computer security and computer networks. She also focused on computer science curriculum and women engineering faculty. Her many awards include being named Fellow of IEEE, SWE and the ACM (Association for Computing Machinery), the IEEE Centennial Medal, the SWE Distinguished Engineering Educator Award, and the Richard E. Merwin Award. She was a significant mentor to women faculty and women in electrical and/or computer engineering and computer science around the world [26–28].

13 Eleanor Baum (1940–)

Eleanor Baum (Fig. 13) has spent her life breaking down barriers. As reported in the *National Enquirer* (!), one of the engineering colleges to which she applied did not accept her—as they had no women's bathrooms! She decided to become an engineer "as a form of rebellion, my way of shocking my parents and teachers" who wanted her to become a high school math teacher.

Electrical engineer Baum was the first female engineering dean in the United States—at The Cooper Union in New York City. When she was an undergraduate student, she was the only female in the engineering class at City College of New York (she graduated in 1959). "I felt very conspicuous," she said, "but I was stubborn

Fig. 13 Eleanor Baum. (Photograph courtesy of Society of Women Engineers Records, Walter P. Reuther Library, Wayne State University)

enough to stick it out." She completed her MS (1961) and PhD (1964) degrees from the Polytechnic Institute of New York.

Baum rose to national and international prominence as a trailblazer in engineering curriculum development and the recruitment of women and minorities to the field. She played leadership roles in a half-dozen professional associations and served on national engineering education committees. Baum was the first female president of the American Society for Engineering Education and also served as the President of ABET.

Baum was very active in recruitment and retention efforts to increase the number of women and minorities in the engineering profession. She served as a role model for women in higher education. Baum was a frequent writer and speaker on this topic and was often a guest on radio and TV programs. In addition, she was quoted as an authority in the field by the press. Baum undertook groundbreaking and widely disseminated national surveys of women engineers and women engineering undergraduates. Those surveys dealt with demographics, attitudes, and personality traits and helped define and direct the national efforts to recruit more women into the engineering profession. She was a frequent speaker on engineering education issues.

She was the 1988 recipient of the Emily Roebling Award presented by the National Women's Hall of Fame. Dr. Baum is a Fellow of the ABET, IEEE, and SWE. She has been inducted into the Women in Technology International Hall of Fame and into the National Women's Hall of Fame. A recipient of SWE's Upward Mobility Award, Baum served on the Boards of Directors of several corporations [29–35].

14 Virginia Sulzberger (1941–)

After receiving her BS (1962) and MS (1966) degrees from the Electrical Engineering Department at the Newark College of Engineering (today the New Jersey Institute of Technology), Virginia Sulzberger worked as a senior planning advisor for Exxon Corporation and in various positions at Public Service Electric & Gas (PSE&G). She worked from 1985 until her retirement in 2006 at the North American Electric Reliability Corporation (NERC). During those years, Sulzberger served as the director of engineering and coordinated planning and engineering activities of NERC's Planning Committee and its many subgroups, which involved more than 200 engineer representatives from the USA and Canada. Sulzberger also developed one of NERC's flagship documents, the long-term reliability assessment. She then served as a consultant for electric power systems.

In 2015, she was elected to the National Academy of Engineering "For leadership and development of electric power system reliability standards." She received the 2014 IEEE Power and Energy Society Lifetime Achievement Award with a citation that reads "For pioneering leadership in developing reliability analysis methods, performing reliability assessments of interconnected transmission systems, and establishing transmission reliability planning standards applied in North America during a 50-year career." On the occasion of this award, Gerry Cauley, president and CEO of NERC said "NERC is pleased that Virginia's many contributions are being recognized by IEEE. Virginia paved the way for NERC's independent reliability assessments of the North American bulk power system and also was instrumental in developing the original NERC planning standards." Her other honors include Fellow in the IEEE, and membership in Eta Kappa Nu and Tau Beta Pi [36–39].

15 Leah Jamieson (1949–)

A member of the Purdue University engineering faculty since 1976, Leah Jamieson's research has focused on speech analysis and recognition, the design and analysis of parallel processing algorithms, and the application of parallel processing to digital speech, image, and signal processing. Her activity in IEEE has been extensive and includes serving as President and CEO of the organization as well as President of the IEEE Foundation. A member of the National Academy of Engineering, she has advocated for women in the STEM fields for many years. Dean Emerita of the College of Engineering at Purdue, the Women in Engineering Program there was named in her honor. Jamieson has received numerous awards over her career including honorary doctorates. She completed her undergraduate mathematics degree at MIT and her master's and PhD in Electrical Engineering and Computer Science from Princeton University [40].

16 Judith Estrin (1954–)

Silicon Valley and networking pioneer Judith (Judy) Estrin was involved in the internet in the early days—working with Vincent Cerf, called one of the fathers of the internet, on Transmission Control Protocol/Internet Protocol (TCP/IP) when she was at Stanford. She has founded eight companies since 1981, served as CTO of Cisco Systems, written a book and served on the boards of directors of companies including FedEx Corporation, Sun Microsystems, and The Walt Disney Company.

Estrin earned her bachelor's degree in mathematics and computer science from UCLA and her master's in electrical engineering in 1977 from Stanford University. She has been named three times to *Fortune* magazine's list of the 50 Most Powerful Women in American Business [41–43].

17 Kristina Johnson (1957–)

Dr. Kristina M. Johnson (Fig. 14) is internationally recognized as an innovator, entrepreneur, and leader in academia and government. She is a pioneer in optoelectronic processing systems, 3D imaging, and color management systems. Her work has formed the basis for 3D films, rear projection systems for televisions, digital mammograms, faster screening of Pap smears for cervical cancer, and foundational work that has resulted in near-to-the-eye displays. Johnson has 45 U.S. patents allowed and pending and is the founder or co-founder of seven successful startup

Fig. 14 Kristina M. Johnson. (Courtesy of Colorado Women's Hall of Fame)

companies. She served as Under Secretary of Energy in the Obama Administration and has championed educating a diverse technical workforce. Johnson received her BS, MS, and PhD degrees in electrical engineering from Stanford and was a post-doctoral fellow at Trinity College Dublin.

Johnson began her career as an assistant professor of electrical and computer engineering at the University of Colorado at Boulder in 1985, where she was a leader in interdisciplinary research on optoelectronics. The first tenure-track female faculty member in the Electrical and Computer Engineering Department, she rose to the rank of professor during her time in Boulder. There she co-founded the National Science Foundation (NSF) Engineering Research Center (ERC) for Optoelectronic Computing Systems.

Her subsequent positions were at Duke University where she was the first woman dean of engineering, Johns Hopkins University where she was the first woman to serve as Provost and Vice President for Academic Affairs, as Under Secretary of the Department of Energy, Chancellor of the State University of New York (SUNY) system, and President of The Ohio State University. Before her appointment at the SUNY system, Dr. Johnson was the co-founder and CEO of Enduring Hydro and its successor company Cube Hydro Partners, companies that acquired and modernized hydroelectric facilities and developed power at unpowered dams.

Her many awards include the 2004 SWE Achievement Award "For outstanding contributions to the fields of optoelectronic processing systems and liquid crystal devices," the International Dennis Gabor Award (first woman), the Fritz Medal from the American Association of Engineering Societies (first woman), induction into the Colorado Women's Hall of Fame and the National Inventors Hall of Fame, election to the National Academy of Engineering, and the IEEE Mildred Dresselhaus Award [44, 45].

18 Ellen Ochoa (1958–)

The first female Hispanic astronaut, electrical engineer Ellen Ochoa (Fig. 15) became the Director of the Johnson Space Center in 2012. Selected for the astronaut program in 1990, she served on her first mission in 1993 aboard the space shuttle *Discovery*. She was in space on four separate occasions, logging more than 1000 h in flight.

Ochoa grew up in California, completing her undergraduate education at San Diego State University, earning a BS in physics and her graduate work in electrical engineering at Stanford University. She investigated optical systems for information processing and received three patents for an optical inspection system, an optical object recognition method, and a method for noise removal in images.

Ochoa is the recipient of numerous honors and awards including NASA's Outstanding Leadership Award and the Harvard Foundation Science Award [46, 47].

Fig. 15 Ellen Ochoa. (Courtesy of Wikipedia)

19 Mary Ann Weitnauer (1962–)

Mary Ann Weitnauer became the first female faculty member of Georgia Tech's School of Electrical and Computer Engineering upon the completion of her doctorate in 1989. Today, the senior associate chair for that school also directs the Smart Antenna Research Laboratory, and her research interests include wireless communication, network time synchronization for multi-hop networks, and millimeter wave communication. Weitnauer is the author or co-author of more than 190 academic papers and holds 23 patents and invention disclosures. Dr. Weitnauer received RCA's Vivian A. Carr Award in 2017 for her achievements in the wireless industry [48].

20 Andrea Goldsmith (1964–)

In 2020, when Dr. Andrea Goldsmith became the first woman to receive the Marconi Prize, Vincent Cerf said:

> Andrea has enabled billions of consumers around the world to enjoy fast and reliable wireless service, as well as applications such as video streaming and autonomous vehicles that require stable network performance. Andrea's personal work and that of the many engineers whom she has mentored have had a global impact on wireless networking.

The prize was awarded for her pioneering contributions to the theory and practice of adaptive wireless communications. It recognizes work that she has conducted for more than 30 years beginning with her PhD research—the theories and engineering techniques that allow wireless networks to adapt to ever-changing configurations of location, signal strength, and interference.

Goldsmith earned her BS degree in electrical engineering at the University of California, Berkley in 1986. She returned to her alma mater to earn her masters and PhD degrees, also in electrical engineering. After working in private industry, she taught at the California Institute of Technology before joining the Stanford University faculty in 1999. In 2020, she became the Dean of Engineering at Princeton University. She is a member of the National Academy of Engineering and the American Academy of Arts and Sciences [49–51].

21 Maha Achour (1964–)

The CEO and founder of Metawave, Maha Achour is now developing radar technology for autonomous vehicles. Her company is building SPEKTRA, an analog beamsteering radar system that can differentiate objects in close proximity, work efficiently in all types of weather conditions, and not lose effectiveness during all types of driving scenarios. Named a Woman of Influence in Silicon Valley in 2020, Achour was founder and CEO of Rayspan which developed and commercialized metamaterials RF systems and antennas. Achour earned her PhD in physics at MIT and earned a masters in electrical engineering, communication theory, and systems from the University of California, San Diego [52–54].

22 Anne Chow (1966–)

Anne Chow was appointed the CEO of AT&T Business in 2019. She is the first woman to be named to this position and the first woman of color to be a CEO in the history of AT&T. Her entire career has been at AT&T, where she started as an engineer and is now the highest ranking Asian American. Her responsibilities include providing three million business customers around the world with fiber and mobile communication solutions.

A first-generation American, with parents who emigrated from Taiwan, Chow was good enough at the piano to be in The Julliard School's pre-college program at age 10. Chow earned her BS and MS degrees in electrical engineering from Cornell as well as her MBA. She sits on the board of directors of the Franklin Covey

Company as well as Girl Scouts of the USA, has written a book, and has received many honors [55, 56].

23 Maryam Rofougaran (1967–)

The co-CEO of Movandi, Maryam Rofougaran is an electrical engineer who was instrumental in starting and building the wireless business at Broadcom. Under her leadership, the business grew to more than $3 billion. Broadcom had acquired Innovent Systems, the company she had co-founded in 1998. Innovent developed low cost Wi-Fi and Bluetooth system-on-chip (SoC) solutions. Rofougaran, who has more than 250 patents, co-founded Movandi to provide solutions in the 5G space.

Rofougaran, who came to the USA from Iran to attend college, earned her BS and MS in electrical and electronics engineering from UCLA. She has been named one of the 50 Most Powerful Women in Technology [57–59].

References

1. Daintith, J., & Gjertsen, D. (Eds.). (1999). *A dictionary of scientists*. Oxford University Press.
2. Ogilvie, M. B. (1993). *Women in science: Antiquity through the nineteenth century, a biographical dictionary with annotated bibliography*. MIT Press.
3. Proffitt, P. (Ed.). (1999). *Notable women scientists*. The Gale Group.
4. Ronan, C. A. (1982). *Science: Its history and development among the world's cultures*. The Hamlyn Publishing Group Limited.
5. Suplee, C. (2000). *Milestones of science*. National Geographic.
6. Shearer, B. F., & Shearer, B. S. (Eds.). (1997). *Notable women in the physical sciences: A biographical dictionary*. Greenwood Press.
7. Arnold, L. B. (1984). *Four lives in science: Women's education in the nineteenth century*. Schocken Books.
8. Bailey, M. J. (1994). *American women in science: A biographical dictionary*. ABC-CLIO.
9. Alic, M. (1986). *Hypatia's heritage: A history of women in science from antiquity through the nineteenth century*. Beacon Press.
10. Morrow, C., & Perl, T. (Eds.). (1998). *Notable women in mathematics: A biographical dictionary*. Greenwood Press.
11. Meade, J. (2009). Ahead of their time. In M. E. Layne (Ed.), *Women in engineering: Pioneers and trailblazers*. ASCE (American Society of Civil Engineers) Press.
12. Hatch, S. E. (2006). *Changing our world: True stories of women engineers*. ASCE (American Society of Civil Engineers) Press.
13. Letizia, A. Bertha Lamme: Pioneering Westinghouse Engineer. Retrieved October 26, 2015, from http://www.geekpittsburgh.com/innovation/bertha-lamme
14. Ogilvie, M., & Harvey, J. (Eds.). (2000). *The biographical dictionary of women in science: Pioneering lives from ancient times to the mid-20th century*. Routledge.
15. Clarke, E. National Inventors Hall of Fame. Retrieved November 30, 2018, from https://www.invent.org/inductees/edith-clarke
16. Goff, A. C. (1946). *Women can be engineers*. Edwards Brothers, Inc.
17. Ingels, M. (1952, September 4). *Petticoats and slide rules*. Western Society of Engineers.

18. Society of Women Engineers, Historical Record of Policy and Interpretation, in Jill Tietjen's possession, approved November 7, 1990.
19. Peden, I. C. Wikipedia. Retrieved July 31, 2024, from https://en.wikipedia.org/wiki/Irene_C._Peden
20. Pope, K. Trailblazing Engineer Irene Peden broke Antarctic barriers for women. *Smithsonian Magazine*. Retrieved June 7, 2019, from https://www.smithsonianmag.com/science-nature/trailblazing-engineer-irene-peden-broke-antarctic-barriers-women-180972330/
21. Rigas, H. B. ETHW. Retrieved July 31, 2024, from https://ethw.org/Harriett_B._Rigas
22. Rigas, H. B. Association of Faculty Women. Washington State University. Retrieved July 31, 2024, from https://afw.wsu.edu/awards/outstanding-students-in-graduate-studies/dr-harriett-b-rigas-2/
23. Rigas, H. B. Wikipedia. Retrieved July 31, 2024, from https://en.wikipedia.org/wiki/Harriett_B._Rigas
24. Conway, L. National Inventors Hall of Fame. Retrieved May 29, 2024, from https://www.invent.org/inductees/lynn-conway
25. Conway, L. Wikipedia. Retrieved May 27, 2024, from https://en.wikipedia.org/wiki/Lynn_Conway
26. Sloan, M. Michigan Tech. Retrieved September 7, 2024, from https://www.mtu.edu/alumni/recognition/profiles/martha-sloan.html
27. Sloan, M. Michigan Technology University, Electrical and Computer. Retrieved August 26, 2024, from https://www.mtu.edu/ece/department/emeriti/sloan/
28. Sloan, M. E. Wikipedia. Retrieved August 26, 2024, from https://en.wikipedia.org/wiki/Martha_E._Sloan
29. Ambrose, S. A., Dunkle, K. L., Lazarus, B. B., Nair, I., & Harkus, D. A. (1997). *Journeys of women in science and engineering: No universal constants* (pp. 422–425). Temple University Press.
30. Baum, E. (1993, February). IEEE Spectrum (pp. 42–44).
31. Baum, E. Women in Technology International (pp. 1–2). Retrieved July 1, 1999, from www.witi.com/center/witimuseum/halloffame/previousinducte/1996/dbaum.shtml
32. Cooper Union. (1998). Why outward bound for engineers? (p. 1). Retrieved December 30, 1998, from www.cooper.edu/engineering/projects/outwardbound/obstory1.html
33. Gale Research. (1995). *Who's who in technology* (7th ed., p. 1328). Gale Research, Inc.
34. The National Enquirer. (1991, June 11). Rebel with a cause becomes 1st lady dean of engineering (p. 10).
35. Tietjen, J. Eleanor Baum. Grandma Got STEM blog. Retrieved July 10, 2013, from https://ggstem.wordpress.com/2013/07/10/eleanor-baum/
36. Business Wire. IEEE Power & Energy Society awards recognize important member contributions at its 2014 general meeting. Retrieved July 30, 2014, from http://www.businesswire.com/news/home/20140729006030/en/IEEE-Power-Energy-Society-Awards-Recognize-Important
37. National Academy of Engineering, National Academy of Engineering Elects 67 Members and 12 Foreign Members. Retrieved February 5, 2015, from https://www.nae.edu/Projects/MediaRoom/20095/130169/130172.aspx
38. NERC. Virginia Sulzberger receives Prestigious IEEE Lifetime Achievement Award. Retrieved April 17, 2014, from http://www.nerc.com/news/Headlines%20DL/Sulzberger%2017APR14.pdf
39. NJIT News Room. NJIT alumna will be inducted into the National Academy of Engineering. Retrieved August 4, 2015, from http://www.njit.edu/news/2015/2015-203.php
40. Jamieson, L. H. (2021, March). Purdue University. Biography. Retrieved April 22, 2021, from https://engineering.purdue.edu/~lhj/biosketch.pdf
41. Estrin, J. JLabs. Retrieved April 22, 2021, from http://www.jlabsllc.com/
42. Estrin, J. Wikipedia. Retrieved April 22, 2021, from https://en.wikipedia.org/wiki/Judith_Estrin
43. Estrin, J. Women in Technology Hall of Fame. Retrieved April 22, 2021, from https://www.witi.com/halloffame/240551/Judy-Estrin-Chief-Executive-Officer-JLABS,-LLC/

44. Johnson, K. M. Wikipedia. Retrieved April 2, 2023, from https://en.wikipedia.org/wiki/Kristina_M._Johnson
45. Unpublished nominations in the files of author Jill S. Tietjen.
46. Biography.com. Ellen Ochoa, Biography: Engineer, Astronaut. Retrieved June 6, 2015, from http://www.biography.com/people/ellen-ochoa-10413023
47. NASA, Johnson Space Center Director Dr. Ellen Ochoa. Retrieved June 6, 2015, from http://www.nasa.gov/centers/johnson/about/people/orgs/bios/ochoa.html
48. Weitnauer, M. A. Georgia Tech. Retrieved April 22, 2021, from https://www.ece.gatech.edu/faculty-staff-directory/mary-ann-weitnauer
49. Goldsmith, A. Wikipedia. Retrieved May 1, 2021, from https://en.wikipedia.org/wiki/Andrea_Goldsmith_(engineer)
50. Schulz, S. (2020, April 15). Andrea Goldsmith, entrepreneur and leader in wireless communications, appointed Princeton University dean of engineering. Retrieved May 1, 2021, from https://www.princeton.edu/news/2020/04/15/andrea-goldsmith-entrepreneur-and-leader-wireless-communications-appointed
51. Stanford News. (2020, April 30). Andrea Goldsmith becomes first woman to win the Marconi Prize, shattering a glass ceiling in the field of telecommunications. Retrieved May 1, 2021, from https://news.stanford.edu/2020/04/30/andrea-goldsmith-first-woman-win-marconi-prize/
52. Achour, M. LinkedIn. Retrieved May 1, 2021, from https://www.linkedin.com/in/mahaachour/
53. Bizjournals. (2020, May 22). A woman of influence Maha Achour develops radar tech for cars. Retrieved May 1, 2021, from https://www.bizjournals.com/sanjose/news/2020/05/22/women-of-influence-maha-achour-metawave.html
54. Metawave. About us. Retrieved May 1, 2021, from https://www.metawave.co/about-us
55. Cornell Chronicle. (2021, March 11). AT&T Business CEO to give technology lecture March 25. Retrieved April 29, 2021, from https://news.cornell.edu/stories/2021/03/att-business-ceo-give-technology-lecture-march-25
56. Murphy, S. H. (2020, April). Meet the Julliard-trained Pianist who became leader of a $37 billion Business Group at AT&T. Retrieved May 1, 2021, from https://www.dmagazine.com/publications/d-ceo/2020/april/meet-the-juilliard-pianist-who-became-leader-of-a-37-billion-business-group-at-att/
57. Movandi. About. Retrieved May 1, 2021, from https://movandi.com/about/
58. Rofougaran, M. LinkedIn. Retrieved May 1, 2021, from https://www.linkedin.com/in/maryam-rofougaran-411aa84/
59. UCLA, Samueli Electrical and Computer Engineering. Maryam Rofougaran: Transforming wireless technology. Retrieved May 1, 2021, from https://www.ee.ucla.edu/maryam-rofougaran-transforming-wireless-technology/

Jill S. Tietjen, PE, entered the University of Virginia in the Fall of 1972 (the third year that women were admitted as undergraduates after a suit was filed in court by women seeking admission) intending to be a mathematics major. But midway through her first semester, she found engineering and made all of the arrangements necessary to transfer. In 1976, she graduated with a BS in Applied Mathematics (minor in Electrical Engineering) (Tau Beta Pi, Virginia Alpha) and went to work in the electric utility industry.

Galvanized by the fact that no one, not even her PhD engineer father, had encouraged her to pursue an engineering education and that only after her graduation did she discover that her degree was not ABET-accredited, she joined the Society of Women Engineers (SWE) and for more than 45 years has worked to encourage young women to pursue science, technology, engineering, and mathematics (STEM) careers. In 1982, she became licensed as a professional engineer in Colorado.

Tietjen started working jigsaw puzzles at age 2 and has always loved to solve problems. She derives tremendous satisfaction seeing the result of her work—the electricity product that is so reliable that most Americans just take its provision for granted. Flying at night and seeing the lights below, she knows that she had a hand in this infrastructure miracle. As an expert witness, she worked to plan, license, and put into rate base new power plants.

Her efforts to nominate women for awards began in SWE and have progressed to her acknowledgment as one of the top nominators of women in the country. Her nominees have received the National Medal of Technology and the Kate Gleason Medal; they have been inducted into the National Women's Hall of Fame and state Halls including Colorado, Maryland and Delaware and have received university and professional society recognition. Tietjen believes that it is imperative to nominate women for awards—for the role modeling and knowledge of women's accomplishments that it provides for the youth of our country.

Tietjen received her MBA from the University of North Carolina at Charlotte. She has been the recipient of many awards including the Distinguished Service Award from SWE (of which she has been named a Fellow and is a National Past President), the Distinguished Alumna Award from the University of Virginia, the Distinguished Alumna Award from the University of North Carolina at Charlotte, and the Distinguished Engineering Alumni Award (special category as a non-alum) from the University of Colorado at Boulder. She has been inducted into the Colorado Women's Hall of Fame and the Colorado Authors' Hall of Fame and elected to the National Academy of Construction. Tietjen sits on the board of Georgia Transmission Corporation and served for 11 years on the board of Merrick & Company.

Her publications include the bestselling and award-winning book *Her Story: A Timeline of the Women Who Changed America* for which she received the Daughters of the American Revolution History Award Medal and *Hollywood: Her Story, An Illustrated History of Women and the Movies* which has received numerous awards. Her award-winning book *Over, Under, Around and Through: How Hall of Famers Surmount Obstacles* was released in 2022. The memoir of Dr. Antonia Novello, *Duty Calls: Lessons Learned from an Unexpected Life of Service*, which she co-wrote, was released in 2024 in English and Spanish-language version. She is the series editor of the Springer Women in Engineering and Science series, has authored two volumes in the series, served as co-author for a few volumes and also as co-volume editor for additional volumes, and written chapters for a number of volumes.

Radomes

Aleksandra Markina-Khusid

1 Introduction

A radome is a structural, weatherproof enclosure that protects a radar antenna. The term "radome" is a portmanteau of "radar" and "dome." These enclosures are designed to be transparent to radio waves in the intended frequency band, allowing the radar system to function effectively while shielding the antenna from environmental factors such as wind, rain, snow, and debris. Radomes are used in various applications, including aviation (covering radar antennas on aircraft), maritime (protecting shipboard radar systems), and ground-based radar installations. Radomes are used in military and civil radars, surveillance systems, telecommunications, weather radars, coastal surveillance, satellite communications, microwave systems, broadcast equipment, and flight simulation systems. Radomes can be constructed from materials such as fiberglass, PTFE-coated fabric, or other composite materials, depending on the specific requirements of the radar system and the environmental conditions it will face. The design and material choice ensure minimal signal attenuation and distortion, maintaining the radar's performance while providing necessary protection to the enclosed antenna [1].

Radomes serve a range of purposes related to the protection and performance of radar systems:

- Environmental protection: Radomes shield radar antennas from harsh environmental conditions such as rain, snow, ice, wind, and debris. This protection ensures the radar system can operate reliably in challenging weather conditions.
- Performance preservation: By protecting the radar antenna from physical damage and environmental interference, radomes help maintain the accuracy and efficiency of the radar system. Maintaining performance is essential for

A. Markina-Khusid (✉)
The MITRE Corporation, Bedford, MA, USA

K. J. Bunker (ed.), *Women in Electrical Engineering*, Women in Engineering and Science, https://doi.org/10.1007/978-3-032-00133-7_2

applications that require high-precision data, such as air traffic control, weather monitoring, and military operations.
- Aerodynamic efficiency: Aviation radomes are designed to be aerodynamically efficient, reducing drag on the aircraft to ensure that antenna installation has a minimal effect on aircraft performance and fuel efficiency.
- Reduced maintenance: By protecting the radar antenna from environmental factors and physical damage, radomes reduce the need for frequent maintenance and repairs, leading to lower operational costs and increased system uptime.
- Signal integrity: Radomes are constructed from materials that are largely transparent to radio waves in the intended frequency band, ensuring minimal signal attenuation and distortion. This transparency is critical for maintaining the integrity and accuracy of the radar signals.
- Stealth and aesthetics: In some military applications, radomes can be designed to reduce the radar cross-section of an aircraft or vehicle, contributing to stealth capabilities. Additionally, radomes can improve the overall aesthetics of a structure or vehicle by concealing the radar antenna.
- Operational safety: In aviation and maritime contexts, radomes contribute to the safety of operations by ensuring that radar systems function correctly, providing essential data for navigation, collision avoidance, and weather monitoring.

Radomes play a crucial role in ensuring the reliability, efficiency, and safety of radar systems across various industries and applications [2]. Figures 1, 2, and 3 illustrate a variety of uses of radomes.

Fig. 1 Haystack radome. (Haystack Radio Telescope—Haystack Observatory—DSC04026—Haystack Observatory—Wikipedia)

Fig. 2 Helicopter radome. (File:JASDF CH-47J(LR) (37-4501) Radome at Komaki Air Base February 23, 2014.JPG—Wikimedia Commons)

Fig. 3 Sea-based X-band radar. (File:Sbx underway.jpg—Wikimedia Commons)

2 Radome Design Considerations

Radome engineering presents a unique set of challenges that designers and manufacturers must navigate to create effective protective enclosures for antenna systems. Radomes serve as structural, weatherproof enclosures that protect radar antennas from environmental factors such as wind, rain, ice, sand, and ultraviolet rays. They also conceal the antenna's electronic equipment from public view and protect nearby personnel from being struck by rotating antennas. Radomes must withstand temperature extremes, moisture and humidity, UV radiation, and erosion, particularly in high-speed applications [1].

The design of radomes involves several key considerations, including material selection, electromagnetic performance, environmental factors, manufacturing and quality control, maintenance and repair, and integration with antenna systems. Material selection is crucial for radome performance, as design must balance electromagnetic transparency, structural strength, weight considerations, and cost-effectiveness. The challenge lies in finding materials that excel in all these areas, often requiring compromises or innovative composite solutions. Common materials used in radome construction include fiberglass and PTFE-coated fabric, which are chosen for their ability to minimally attenuate electromagnetic signals [3].

Electromagnetic performance is a primary concern in radome design. The radome should introduce minimal signal loss and minimize internal reflections that can cause signal distortions [4]. Integration with antenna systems is another important consideration. Radomes must be designed in concert with the antenna systems they protect, ensuring minimal impact on the antenna's radiation pattern and accounting for boresight error and near-field effects [5]. This requires a multidisciplinary approach, combining expertise in materials science, electromagnetic theory, structural engineering, and manufacturing processes [6].

Advanced composites and special products are made using reinforcements like fiberglass, quartz, graphite, and Kevlar, along with materials such as polyester, epoxies, and cyanate ester. Core materials like honeycomb and foams are also used, with parts cured at high temperatures and pressures depending on the application. The design of a radome involves calculating transmission efficiency using plane wave solutions and taking into account dielectric constant and sheet thickness of each layer, wave frequency, polarization, and incident angle.

Radomes for high-speed applications, such as missiles, face challenges like ultra-high temperatures. Silicon nitride ceramic is a preferred material due to its superior properties. Broadband radome wall structures like A-Sandwich, C-Sandwich, and multilayer are commonly used. The design of metamaterial radomes involves determining electromagnetic characteristics around the operating frequency. These radomes are made by aligning layers of square Split-Ring Resonators (SRRs) with air layers, optimizing properties like negative permeability with low losses.

The electromagnetic performance of sandwich radomes is affected by panel window areas and flanges. Flanges must be impedance-matched to the window area to minimize loss and phase shift. Two common flange frameworks are perpendicular joints and parallel lap joints, with the former having better performance due to a narrower cross-section. Hydrophobic coatings are used to reduce signal attenuation caused by water, which has a high dielectric constant and loss tangent at microwave frequencies. These coatings cause water to bead up and run off the radome surface [3].

3 Materials Used in Radomes

Radome materials are carefully selected to balance electromagnetic transparency, structural integrity, and resilience to environmental conditions. A combination of factors including operating frequency, environmental and structural requirements, and cost constraints, often result in custom-engineered solutions for specific applications [1]. Main types of materials used in radome design are summarized in Table 1.

3.1 Traditional Materials

Fiberglass is one of the most widely used materials in radome manufacturing. It offers excellent electromagnetic transparency, low dielectric constant, and good structural properties. E-glass and S-glass are popular fiberglass variants, with S-glass providing superior strength-to-weight ratio.

Quartz fiber is increasingly used in high-performance radomes, especially for aerospace applications. It offers better RF transparency than fiberglass and maintains its properties at higher temperatures, making it ideal for supersonic aircraft.

Aramid fibers, such as Kevlar, are employed for their high strength-to-weight ratio and impact resistance. They are often used in sandwich constructions or as reinforcement in composite structures.

Polyester and epoxy resins are the primary matrix materials used to bind the fibers together. Epoxy resins generally offer better mechanical properties and environmental resistance compared to polyester resins. Stretched and laminated resin fiber materials are a common choice for single-wall radomes.

For core materials in sandwich constructions, foam and honeycomb structures are common. These include polymethacrylimide (PMI) foam, polyvinyl chloride (PVC) foam, and aramid or aluminum honeycomb. These materials provide structural support while maintaining low weight and minimal RF interference [7].

3.2 Advanced Polymers

Advanced thermoplastics have replaced initial pure polymer-based monolithic thermoplastic radomes due to their superior resistance to rain erosion, moisture absorption, and impact loads. These advanced polymers offer improved mechanical properties and lower coefficients of thermal expansion (CTE), making them suitable for various radome applications.

Advanced polymer composites, such as cyanate ester resin systems, are promising for bridging the technology gap between 150 and 500 °C. Cyanate ester resins cure by cyclotrimerization to form polycyanurate containing triazine rings, offering

Table 1 Radome materials

Material type	Material	Key properties	Applications
Traditional materials	Fiberglass (E-glass, S-glass)	Excellent electromagnetic transparency, low dielectric constant, good structural properties	Widely used in various radome applications
	Quartz fiber	Superior RF transparency, maintains properties at high temperatures	Aerospace applications, supersonic aircraft
	Aramid fibers (Kevlar)	High strength-to-weight ratio, impact resistance	Sandwich constructions, reinforcement in composites
	Polyester and epoxy resins	Good mechanical properties, environmental resistance	Binding fibers in composite structures
	Foam and honeycomb structures	Low weight, minimal RF interference	Core materials in sandwich constructions
Advanced polymers	Advanced thermoplastics	Superior resistance to rain erosion, moisture absorption, impact loads	Various radome applications
	Cyanate ester resins	Balanced thermal and dielectric properties, lightweight	Supersonic airborne platforms
Ceramic materials	Silicon nitride	High mechanical strength, thermal stability, low dielectric constant	High-temperature radomes, tactical missile applications
	Alumina	High-temperature resistance, but high dielectric constant	Specialized high-temperature applications
	Pyroceram	Better thermal shock resistance than alumina	High-speed applications
	Fused silica	Low thermal expansion, excellent thermal shock resistance	Limited by high porosity and low mechanical strength
	Boron nitride	Good electrical properties, limited by poor mechanical properties	Specialized applications
Emerging materials	Metamaterials	Engineered electromagnetic properties, potential for enhanced RF performance	Next-generation radomes, high-performance applications
Special-purpose layers	Hydrophobic coatings	Prevents water accumulation, fluoropolymer-based	Enhances signal transmission in wet conditions
	Metallic strips/conductive coatings	Lightning protection, minimizes electromagnetic interference	Integrated into radome design

balanced thermal and dielectric properties. These advanced polymers provide lightweight, high-performance radomes suitable for supersonic airborne platforms, addressing the need for materials that can withstand higher temperatures while maintaining desirable mechanical and dielectric properties [8].

3.3 *Ceramic Materials*

Ceramics are materials of choice in specialized applications where extreme heat resistance is required, such as in high-speed aerospace technology. Ceramics are valued for their superior thermal and mechanical properties. Key ceramics used in radome fabrication include silicon nitride, alumina, pyroceram, fused silica, and boron nitride [8].

Silicon nitride stands out for its high mechanical strength, thermal stability, and low dielectric constant, making it ideal for high-temperature radomes, especially in tactical missile applications [9]. It is suitable for flight speeds over Mach 7 due to its superior properties compared to other materials like fused silica and pyroceramics. Fabrication methods such as slip casting and gel casting are explored to optimize its properties for radome use.

Alumina is noted for its high-temperature resistance, withstanding temperatures above 1760 °C, but it has drawbacks like a high dielectric constant and low thermal shock resistance. Pyroceram offers better thermal shock resistance than alumina but is less stable dielectrically at higher speeds. Fused silica is valued for its low thermal expansion and excellent thermal shock resistance, although its high porosity and low mechanical strength limit its application in high-speed radomes. Boron nitride, while having good electrical properties, is limited by poor mechanical properties [10].

3.4 *Emerging Materials*

Emerging materials and metamaterials are being researched for their potential to enhance radome performance and provide multi-functional capabilities in next-generation radomes. These materials are explored for their ability to improve RF performance and address the limitations of traditional materials used in radome construction, particularly for applications involving extreme conditions such as hypersonic speeds and high temperatures.

Metamaterials are engineered to have specific electromagnetic properties that can overcome the limitations of conventional ceramics and polymers. They offer the potential to enhance antenna gain, reduce boresight error, and improve overall radome performance. Designs such as split-ring resonators (SRR) and fishnet structures are examples of metamaterial configurations that show promise in enhancing radome capabilities. Preliminary experimental results have demonstrated the potential of high-temperature metamaterials. For instance, a high-temperature SRR structure fabricated using an alumina-silicon substrate and carbon trace has shown promising dielectric properties at elevated temperatures, indicating the feasibility of using metamaterials in high-performance radome applications [11].

3.5 *Special-Purpose Layers and Materials*

Hydrophobic coatings are applied to radome surfaces to prevent water accumulation, which can interfere with signal transmission. These coatings are typically fluoropolymer based.

For lightning protection, metallic strips or conductive coatings are sometimes incorporated into the radome design, carefully positioned to minimize electromagnetic interference.

4 Radome Architectures

Radomes can be classified based on their construction and material properties. Each class has its own advantages and is chosen based on factors like the operational environment, frequency range, and structural requirements. Radomes are also characterized by frequency bands, operational bandwidth, and efficiency. This section describes are the main classes based on wall architecture.

4.1 *Monolithic Radomes*

Monolithic radomes are a type of radome characterized by their solid wall construction, typically made from a single material. These structures are designed to protect antennas from environmental elements while minimizing interference with radio waves. Monolithic radomes can be categorized into two main types based on their wall thickness: half-wave and electrically thin-walled radomes [2].

Monolithic radomes are constructed using materials that are transparent to radio waves, such as fiberglass or other composite materials. The design often involves a single, continuous layer, which can be enhanced with additional lattice structures for improved performance. For example, a monolithic wideband millimeter-wave radome may include multiple solid and lattice layers formed from the same material, with the lattice layers featuring void regions created by selectively omitting material during construction. This design helps in maintaining structural integrity while allowing for effective transmission of electromagnetic signals.

Monolithic radomes are widely used in various applications due to their simplicity and effectiveness. They are commonly employed in aerospace and defense industries to protect radar systems on aircraft and ground-based installations. The solid wall construction provides durability and resistance to harsh environmental conditions, such as wind, rain, and UV exposure. Additionally, these radomes do not introduce significant boresight errors, making them suitable for applications requiring high precision.

The performance of monolithic radomes is influenced by their material properties and design features. They are designed to minimize transmission loss and reflection effects on electromagnetic signals. For instance, a monolithic radome made of glass composite may exhibit a transmission loss of 0.7 dB, which is relatively low and acceptable for many radar applications. The use of advanced materials and coatings can further enhance their electromagnetic performance across various frequency bands. Monolithic radomes offer a robust solution for protecting sensitive radar equipment while ensuring minimal impact on signal quality. Their straightforward design and construction make them a cost-effective choice for many industrial applications [2].

4.2 Sandwich Radomes

Sandwich radomes are a sophisticated type of radome design characterized by their multilayered wall construction, as illustrated in Fig. 4. These structures are engineered to provide superior electromagnetic performance while maintaining structural integrity and environmental protection for antenna systems. Sandwich radomes are designed to perform well over specific frequency bands, with the ability to be optimized for narrow bandwidths or multiple discrete frequencies [12].

Sandwich radome consists of three or five primary layers:

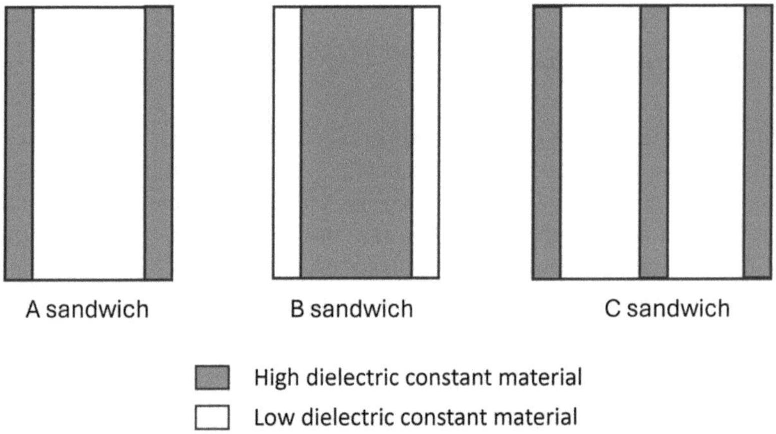

Fig. 4 Sandwich Radome types

- **A-sandwich**: Two high-density polymer skins (high dielectric constant) sandwiching a low-density core material (low dielectric constant).
- **B-sandwich**: Two layers of low dielectric constant materials sandwiching a high dielectric constant material in the middle.
- **C-sandwich**: Three layers of skin and two core layers.

Common high dielectric constant materials include fiberglass-epoxy composites, quartz-cyanate ester composites, and aramid fibers such as Kevlar, all of which provide the necessary structural integrity and electromagnetic performance. Low dielectric constant material options include various foam cores like polyurethane, polymethacrylimide (PMI), and polyvinyl chloride (PVC) foams. Honeycomb structures, such as aramid and aluminum honeycombs, are also popular choices due to their excellent strength-to-weight ratios. Additionally, wood cores may be used in specific underwater applications, while low-density polymer materials offer versatility for various designs. These material choices enable the construction of radomes that meet the specific performance requirements of different applications [13].

Advantages of sandwich radomes include self-supporting structure, eliminating the need for air pressurization; reduced component count, leading to lower construction and maintenance costs; customizable performance through adjustments in wall thickness and material properties; excellent durability; and weather resistance. Sandwich radomes are widely used in various applications, including sophisticated 3D air surveillance radar systems, military and commercial satellite communications, civilian air traffic control, weather radar, and phased-array radar systems.

Sandwich radomes can be manufactured in a wide range of sizes and are designed to withstand high wind speeds. Sandwich radomes represent a highly effective solution for protecting sensitive radar and communication equipment while ensuring optimal electromagnetic performance. Their versatility, durability, and customizable nature make them a popular choice for a wide range of applications in both military and civilian sectors [14].

4.3 Air-Filled Radomes

Air-filled radomes, also known as air-supported or inflatable radomes, are a specialized type of protective enclosure for antenna systems that utilize air pressure to maintain their shape and structural integrity. These innovative structures offer a unique blend of functionality, flexibility, and cost-effectiveness for various applications in telecommunications, military operations, and scientific research [15].

Air-filled radomes are constructed using high-strength, lightweight fabrics such as polyester, PVC, TPU (thermoplastic polyurethane), or fiberglass-reinforced polymer composites. The outer surface is often coated with a weather-resistant and radio-transparent membrane, like fluoropolymer or silicone elastomer, to enhance durability and signal transmission. The radome's structure is maintained by continuous air pressure, creating a stable, self-supporting enclosure around the antenna equipment. This design eliminates the need for rigid internal frameworks, resulting in a lightweight and portable solution.

Air-filled radomes boast lightweight construction: they are significantly lighter than conventional radomes, facilitating easy transport and installation, especially in remote or challenging locations. Air-filled radomes have lower manufacturing and installation costs compared to traditional solid radomes. Despite their lightweight design, air-filled radomes offer excellent durability against harsh weather conditions. Air-filled radomes are versatile and scalable; and they can be customized to accommodate various antenna sizes and configurations. Finally, they can be rapidly deployed, minimizing downtime in critical operations.

Air-filled radomes are used in mobile telecommunications, especially for temporary or remote deployments; military operations requiring rapidly deployable communication systems; scientific research and remote sensing projects; emergency response and disaster relief communications.

While air-filled radomes offer numerous benefits, they require continuous air pressure maintenance and may be more susceptible to punctures or air leaks compared to solid radomes. However, their advantages in terms of portability, cost, and rapid deployment make them an attractive option for many applications where flexibility and quick setup are important [15].

4.4 Membrane Radomes

Membrane radomes are a type of thin-walled radome design characterized by their lightweight construction and minimal material usage. These structures are engineered to provide essential protection for antenna systems while maintaining optimal signal transmission [16].

Membrane radomes typically consist of a single, thin layer of dielectric material, usually less than 0.040 in. (1 mm) thick. Common materials used include fiberglass-reinforced composites, high-performance fabrics like Kevlar, and advanced polymers with low dielectric constants. The thin membrane is supported by a framework of panel flanges that carry the structural loads. This design allows for a minimal amount of material in the signal path, reducing electromagnetic interference, distortion, and loss. Membrane radomes offer advantages of low weight and cost. However, they may have limitations in terms of structural strength compared to thicker designs.

Membrane radomes are typically used in small aircraft and UAVs, portable or temporary antenna installations, and situations where weight savings are crucial. The installation of membrane radomes often involves careful tensioning of the material to ensure proper shape and performance. Regular inspections are necessary to check for any damage or degradation of the thin membrane material.

While membrane radomes excel in providing a lightweight, low-interference solution for antenna protection, they may not be suitable for all applications, especially those requiring high structural strength or extreme environmental resistance.

In such cases, alternative radome designs like solid laminates or sandwich constructions might be more appropriate. Membrane radomes represent an efficient solution for protecting antenna systems in applications where minimal weight and signal interference are primary concerns.

4.5 Metal Space Frame Radomes

Metal space frame radomes utilize a metal framework combined with electromagnetically transparent membranes. These structures offer a unique balance of strength, durability, and RF performance. Metal space frame radomes typically consist of a structural framework made of aluminum alloy such as 6061 T6 chosen for its structural strength and corrosion resistance and thin, electromagnetically transparent membranes bonded to the framework. The framework is usually constructed as a geodesic dome, composed of triangular panels bolted together. For smaller radomes (up to 30 ft in diameter), panels can be uniform or varied in size and shape. Larger radomes often use randomized geometries for improved structural integrity. The most common membrane material is ESSCOLAM™, a proprietary plastics laminate developed specifically for radome applications. Other options include GORE-TEX® and Teflon®-coated fiberglass [17].

Metal space frame radomes are appropriate for a variety of applications, including satellite communications across all military and commercial frequency bands, intelligence gathering, radio astronomy, weather radar, and 2D surveillance radar. The radomes are designed to withstand standard wind speeds of 150 mph, with options for up to 250 mph. They are easy to install, feature hydrophobic coatings for enhanced performance in rain, and include an electrostatic cage for lightning protection [17].

5 Analysis of Radome-Enclosed Antennas

Radome-enclosed antennas are analyzed using several methods and techniques to evaluate their performance and impact on the antenna system [14]. When analyzing radome-enclosed antennas, engineers focus on several critical parameters to assess the system's overall performance and the radome's impact on antenna functionality. These key analysis parameters include:

- Insertion loss and transmission efficiency, which quantify the attenuation of electromagnetic signals as they pass through the radome
- Boresight error and beam deflection, which measure the radome's effect on the antenna's main beam direction and accuracy
- Sidelobe level changes, which indicate how the radome alters the antenna's radiation pattern outside the main beam

- Polarization effects, which assess the radome's influence on the antenna's polarization characteristics
- Bandwidth performance, which evaluates the system's ability to maintain consistent performance across the intended frequency range

Figure 5 illustrates attenuation and distortion of antenna beam caused by a radome.

By carefully examining effects of radome wall construction and overall shape on electromagnetic waves at frequencies of interest, engineers can optimize radome designs to minimize their impact on antenna performance while ensuring adequate protection and aerodynamic properties [18]. Below are some key computational approaches in radome modeling.

5.1 Method of Moments (MoM) with Multilevel Fast Multipole Algorithm (MLFMM)

The Method of Moments (MoM) with Multilevel Fast Multipole Algorithm (MLFMM) is a powerful computational technique used for analyzing electrically large radomes and complex antenna systems. This full-wave electromagnetic analysis method combines the accuracy of MoM with the computational efficiency of MLFMM, making it particularly suitable for modeling intricate radome geometries and their interactions with enclosed antennas. The MLFMM accelerates the matrix-vector multiplication in the iterative solution process of MoM, reducing the

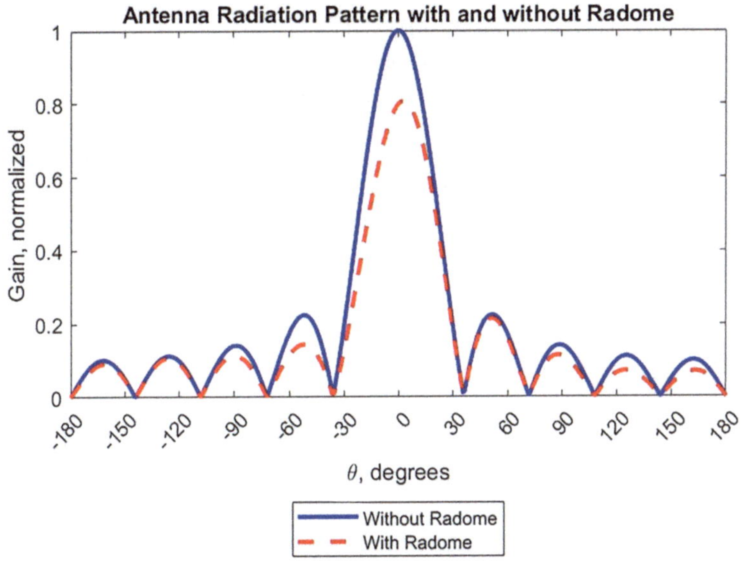

Fig. 5 Antenna beam distortion caused by a Radome

computational complexity from $O(N^2)$ to $O(N \log N)$, where N is the number of unknowns. This significant reduction in computational resources allows for the analysis of much larger structures than traditional MoM alone. The method can handle multilayered radomes and frequency selective surfaces (FSS), providing accurate results for both near-field and far-field characteristics. By using this approach, engineers can efficiently model the electromagnetic behavior of radome-enclosed antennas, including effects such as transmission loss, beam deflection, and polarization changes, while accounting for complex material properties and geometrical features of modern radome designs [19, 20].

5.2 *Finite Element Method (FEM) and Finite-Difference Time-Domain (FDTD)*

The Finite Element Method (FEM) and Finite-Difference Time-Domain (FDTD) are powerful numerical techniques widely used in radome analysis, particularly for structures with complex material properties. These methods discretize the problem domain into small elements or cells, allowing for detailed modeling of intricate radome geometries and material compositions. FEM excels in handling irregular shapes and is especially adept at modeling inhomogeneous and anisotropic materials, making it ideal for analyzing multilayered radomes or those with spatially varying properties. It formulates the problem in terms of variational principles, solving for field quantities at nodal points. FDTD, on the other hand, directly solves Maxwell's equations in the time domain, making it particularly useful for broadband analysis and transient problems. It divides space and time into a grid of cells and steps, respectively, updating electric and magnetic fields alternately. Both methods are capable of providing detailed near-field analysis, which is crucial for understanding field distributions within and around the radome structure. They can accurately model the interactions between the antenna and the radome, including effects such as internal reflections and surface currents. While computationally intensive, especially for electrically large structures, these methods offer unparalleled flexibility in modeling complex material properties and geometries, making them indispensable tools in modern radome engineering [21, 22].

5.3 *Ray Tracing Techniques*

Ray tracing methods are widely employed for analyzing electrically large radomes due to their computational efficiency and ability to handle complex geometries at high frequencies. These techniques are fundamentally based on the principles of geometrical optics, which approximates electromagnetic wave propagation as rays traveling along straight lines. This approach is particularly effective when the wavelength is much smaller than the radome dimensions, a condition typically met in high-frequency applications. Ray tracing algorithms can efficiently model the interactions between electromagnetic waves and the radome structure, including multiple reflections and refractions within radome walls. The method typically involves launching a large number of rays from the antenna source and tracing their paths as they interact with the radome surfaces. By employing techniques such as the Shooting and Bouncing Rays (SBR) method or the Method of Images, ray tracing can account for complex phenomena like internal reflections, which are crucial for accurately predicting radome performance. Advanced ray tracing techniques may also incorporate diffraction effects using the Uniform Theory of Diffraction (UTD), further enhancing the accuracy of the analysis. While ray tracing methods may not capture all the intricacies of full-wave solutions, their computational efficiency makes them invaluable for rapid design iterations and optimization of electrically large radomes, especially in scenarios where full-wave methods would be prohibitively expensive in terms of computational resources [1].

5.4 Hybrid Techniques

Hybrid techniques in radome analysis combine different computational methods to leverage the strengths of each approach while mitigating their individual limitations. These techniques are particularly valuable for analyzing complex radome-antenna systems that span a wide range of electrical sizes and material properties. Domain decomposition is a key strategy in hybrid methods, where the problem is divided into separate regions that are analyzed using the most appropriate technique for each domain. For instance, the antenna and its immediate vicinity might be modeled using a full-wave method like the Method of Moments (MoM) or Finite Element Method (FEM) to capture detailed near-field interactions, while the larger radome structure could be analyzed using asymptotic methods like Physical Optics (PO) or ray tracing. Iterative techniques are often employed to combine these different methods, allowing for the exchange of information between domains to account for mutual coupling effects. For example, the PO-MoM hybrid method uses PO to efficiently model large radome surfaces and MoM for the antenna and radome tip, iteratively updating the solution to capture interactions between the two regions. Another approach is the hybridization of full-wave and asymptotic methods, such as combining FEM with ray tracing, to balance accuracy and computational efficiency. These hybrid techniques enable the analysis of electrically large radomes with complex geometries and material properties, providing more accurate results than purely asymptotic methods while being computationally more efficient than full-wave analysis of the entire structure [23–25].

By combining these analysis techniques, radome engineers can accurately predict and optimize the performance of radome-enclosed antenna systems, balancing electromagnetic transparency with structural and environmental requirements. We recommend [26] for a thorough review of computational methods for radome electromagnetic evaluation.

6 Radome Testing

Accurate radome measurements are essential for assessing their impact on electromagnetic wave transmission and ensuring minimal signal distortion. Experimental methods complement and validate computational techniques while providing essential real-world performance data.

For electromagnetic measurements, experimental methods involve a combination of near-field and far-field measurements to comprehensively characterize radome-antenna systems. Near-field measurements are particularly valuable for assessing radome transmission properties, allowing engineers to map the electromagnetic field distribution close to the radome surface. This data can be used to evaluate insertion loss, phase errors, and local field distortions introduced by the radome. Advanced near-field scanning techniques, often employing robotic positioning systems, enable high-resolution mapping of field patterns [1]. Far-field pattern measurements, on the other hand, provide insight into the overall system performance, including parameters such as gain, beamwidth, sidelobe levels, and boresight error. These measurements are typically conducted in anechoic chambers, such as shown in Fig. 6, or outdoor ranges, using precision positioners to rotate the radome-antenna assembly. Additionally, advanced measurement techniques like near-field to far-field transformations and phase-less measurement methods are being developed to overcome limitations in traditional testing approaches, particularly for electrically large radomes. By combining these experimental methods with computational analysis, engineers can achieve a comprehensive understanding of radome performance, validate theoretical models, and refine designs for optimal real-world operation [14].

Structural measurements include mechanical and environmental testing aspects. Specialized test facilities are crucial for simulating a range of environmental and operational conditions that radomes may encounter. These facilities often include climate chambers, wind tunnels, and impact testing rigs, all designed to evaluate radome performance under realistic scenarios. By replicating conditions such as temperature extremes, humidity, rain, and aerodynamic loads, these facilities provide a comprehensive assessment of radome durability and functionality.

Mechanical testing is a key component of structural measurements, focusing on the radome's ability to maintain its structural integrity under various environmental stresses. Techniques such as tensile testing, impact testing, and fatigue testing are utilized to assess the material's strength and durability. These tests are essential to ensure that the radome can endure physical stresses without compromising its electromagnetic properties.

Environmental testing complements mechanical testing by simulating the diverse conditions a radome might face in real-world applications. This includes tests for

Fig. 6 An electronic technician monitors radome testing. (402nd EMXG: Making sure radomes provide clear site)

thermal cycling, humidity exposure, and UV radiation. The primary objective of environmental testing is to verify that the radome can sustain its performance across a wide spectrum of environmental conditions, ensuring reliability and longevity in its operational environment [1].

Two military standards are applicable to testing of radomes intended for military applications: MIL-STD-810, Department of Defense Test Method Standard—Environmental Engineering Considerations and Laboratory Tests [27], and MIL-STD-461, Department of Defense Interface Standard—Requirements for the Control of Electromagnetic Interference Characteristics of Subsystems and Equipment [28].

MIL-STD-810 provides a framework for assessing how radomes withstand various environmental stresses. By simulating conditions such as extreme temperatures, humidity, rain, and sand, MIL-STD-810 ensures that radomes can endure the harsh environments they may encounter in real-world scenarios. This rigorous testing is essential for verifying that the radome will not compromise the radar system's functionality under adverse conditions [27].

MIL-STD-461 focuses on the control of electromagnetic interference (EMI) characteristics. This standard is applicable to radome testing as it addresses the electromagnetic compatibility of the radome with the radar system it houses. MIL-STD-461 outlines the procedures for testing and measuring the radome's ability to prevent unwanted electromagnetic interference that could disrupt the radar's operation [28].

By adhering to these two standards, manufacturers can ensure that the radome meets the stringent requirements necessary for military applications, ensuring that they perform reliably in the field.

7 Design Optimization of Radomes

Radome optimization is a complex multidisciplinary process that aims to balance various design goals, primarily minimizing the radome's impact on antenna performance while maintaining structural integrity under required environmental conditions. Technical performance metrics considered in an optimization process can include insertion loss, transmission efficiency, boresight error, beam deflection, sidelobe level changes, polarization effects, and bandwidth performance. Sandwich radome designs, in particular, lend themselves well to optimization due to the ability to adjust layer materials and thickness [29].

One novel approach to radome optimization is the use of variable thickness radomes (VTR). VTRs improve electromagnetic performance by reducing phase distortion, a common limitation of constant thickness radomes. By tailoring the thickness profile across the radome surface, designers can minimize insertion loss and boresight error while maximizing power transmission. Multi-objective algorithms, such as particle swarm optimization, have been employed to address multiple performance criteria simultaneously, effectively minimizing both boresight error and transmission loss in airborne radomes [29].

Material selection and layering strategies are crucial in radome optimization. Inhomogeneous radome walls with graded dielectric properties have shown promise in improving broadband performance. By varying the porosity or composition of materials across layers, designers can create radome structures that maintain high power transmission efficiency across a wide frequency range. Some optimized designs have achieved power transmission efficiencies exceeding 99% at specific frequencies [30]. Robust optimization approaches also consider manufacturing tolerances and environmental factors, such as ablation effects on high-speed aircraft radomes, which can significantly alter electromagnetic performance over time [31].

Deng and Zhou [32] present a multidisciplinary optimization design method for composite airborne radomes, aiming to enhance design efficiency and reduce production costs by addressing both structural and electromagnetic performance requirements. The method integrates the aperture integration-surface integration

(AI-SI) algorithm for electromagnetic analysis and the finite element method for mechanical behavior evaluation. The equivalent transmission line (ETL) method is introduced to evaluate the power transmittance coefficient of multilayer structures. The optimization problem, characterized by its nonlinear and multi-peak nature, is tackled using a multi-island genetic algorithm, which is more effective than conventional gradient-based algorithms [32].

Kolak's patent addresses the issue of electromagnetic wave distortion caused by radomes, which can alter wave polarization and lead to degraded performance of communications systems [33]. The invention introduces a method for optimizing the antenna's orientation to counteract distortion without altering the radome's physical design. By determining the optimal angle to rotate the antenna based on axial ratio values, the method reduces wave distortion, significantly improving signal quality and increasing throughput in satellite communication systems [33].

An integrated methodology for optimizing the design of airborne satellite communication (SATCOM) terminals on unmanned aerial vehicles (UAVs) is explored by Markina-Khusid et al. [6]. The focus is on minimizing electromagnetic loss through the radome. The study employs a multidisciplinary model to design the antenna and radome simultaneously, addressing both electromagnetic performance and radome weight. Large airborne radome similar to that on the GlobalHawk aircraft (Fig. 7) was considered in this study. The optimization process involves a C-sandwich radome wall structure and utilizes both gradient-based and heuristic optimization techniques. The study reveals trade-offs between RF loss and radome weight, illustrated by a Pareto front, and highlights the potential for future work to explore more complex antenna designs [6].

Radome optimization is a multifaceted process that requires balancing electromagnetic performance with structural integrity and environmental considerations. The approaches discussed above demonstrate the potential for significant improvements in radome design through innovative techniques and multidisciplinary optimization methods. By integrating various strategies, such as variable thickness

Fig. 7 GlobalHawk UAV. (File:Global Hawk 1.jpg—Wikimedia Commons)

profiles, material selection, and antenna orientation optimization, designers can achieve enhanced performance and efficiency in radome applications.

8 Challenges and Trends in Radome Engineering

As radar and communication technologies continue to advance, the design of radomes must evolve to meet new demands. Manufacturing and quality control of radomes also present significant challenges. Achieving consistent quality requires precision manufacturing to maintain tight tolerances, particularly in sandwich constructions or thin membranes. The construction process demands expertise in materials science, structural engineering, and electromagnetics. Quality assurance is crucial to ensure uniform dielectric properties and structural integrity across the radome.

Maintenance and repair are vital for keeping radomes in optimal condition over time. Detecting damage, such as small defects like pinholes, can be difficult but is essential, as these can significantly impact performance. Repair techniques must restore both structural and electromagnetic properties without introducing new issues. Field maintenance involves developing repair methods that can be applied in various operational environments. Common damages include holes from static discharge, dents, and scratches from impacts. Moisture ingress can cause internal delamination and distort radiation patterns. Proper maintenance is crucial to prevent and repair such damages, restoring the radome to its original condition while considering factors like transmissivity, reflection, and diffraction.

Radome engineering faces several new and emerging challenges as technology advances and requirements become more demanding. One key area of focus is the development of advanced materials and manufacturing techniques to enhance radome performance. The evolution of composite material technology is driving innovation, with new composites being developed to improve properties like transmissivity and durability. Techniques such as vacuum-assisted resin transfer molding (VARTM) are emerging, and the use of quartz fiber in aerospace radomes is increasing for improved performance.

Electromagnetic design challenges are also prominent as radar and communication systems become more advanced. One significant area of focus in radome design is the adaptation to higher frequencies, such as those used in millimeter-wave and terahertz applications, which require materials with extremely precise properties and manufacturing techniques.

There is also a need for multi-band radomes that can operate across wider frequency ranges, and stealth radome designs aim to reduce radar cross-section. Wider bandwidths are required to accommodate modern Active Electronically Scanned Arrays (AESA) radar systems and optimizing electromagnetic transparency while maintaining structural integrity remains an ongoing challenge. Additionally, the advent of AESA presents new challenges in managing heat and ensuring optimal electromagnetic performance. The integration of multiple antenna systems under a single radome further complicates the design process, necessitating the development of multi-function radomes.

Size and weight reduction are constant pressures in radome design. As radar arrays become smaller and more powerful, radomes must shrink accordingly. Weight reduction is a key focus, while still maintaining damage resistance. Engineers must balance size and weight constraints with electromagnetic and structural performance.

Environmental and operational factors also pose engineering challenges for radomes. They must withstand harsh conditions, including damage from hail, ground equipment, foreign objects, and lightning. Higher operating temperatures necessitate new materials and designs, and impact resistance and durability must be improved without compromising performance.

Adapting military radome technology for commercial use presents unique challenges. Commercial aircraft are transitioning to more advanced antenna systems, requiring radome redesigns. Cost constraints in commercial applications limit material and design choices, and regulatory requirements from bodies like the FAA and EASA add complexity to material qualification.

Finally, simulation and optimization tools are becoming crucial for radome engineering. Electromagnetic simulation software helps optimize designs without extensive physical testing, and optimization tools allow engineers to evaluate millions of material and orientation combinations, enhancing the efficiency and effectiveness of the design process.

References

1. Kozakoff, D. J. (1997). *Analysis of radome-enclosed antennas*. Artech House.
2. Gampala, G., Vogel, M., & Reddy, C. J. (2015, May 13). Efficient design and analysis of airborne radomes. Retrieved from https://www.microwavejournal.com/articles/24227-efficient-design-and-analysis-of-airborne-radomes
3. Soumya, V. M., Navaneetha, S., Reddy, A. N., & Kumar, J. J. (2017). Design considerations of radomes: A review. *Journal of Mechanical Engineering and Technology, 8*(3), 42–48.
4. Sheret, T., Parini, C., & Allen, B. (2016). Efficient design of a radome for minimised transmission loss. *IET Microwaves, Antennas & Propagation, 10*(15), 1662–1666.
5. Parameswaran, A., & Sonalikar, H. (2020). Design of airborne radome using novel temperature dependent electromagnetic modeling. *Progress in Electromagnetic Research C, 104*, 37–52.
6. Markina-Khusid, A., Ning, K., Yu, B. Y., & Kolak, F. (2017). Design optimization of a satellite communication terminal: Analyzing SATCOM performance on an unmanned aerial vehicle. *IEEE Antennas and Propagation Magazine, 59*(3), 60–68.
7. Mani, G. S. (2013). Radome materials. In V. R. Murthy, S. Sundaram, & B. Viswanathan (Eds.), *Microwave materials* (1st ed., pp. 200–239). Springer.
8. Kumar, S., & Gupta, P. (2019). A review on ceramic and polymer materials for radome applications. In *IEEE Indian Conference on Antennas and Propagation (InCAP)* (pp. 1–4). IEEE.
9. Chandra, M. P., Ayaz, M. A., Goud, P. N., & Sateesh, N. (2022). Modelling and analysis of aircraft radome using different materials. *Materials Today: Proceedings, 62*, 4492–4497.
10. Saeedi Heydari, M., Ghezavati, J., Abbasgholipour, M., & Mohammadi Alasti, B. (2017). Various types of ceramics used in radomes: A review. *Scientia Iranica, 24*(3), 1136–1147.

11. Kenion, T., Yang, N., & Xu, C. (2022). Dielectric and mechanical properties of hypersonic radome materials and metamaterial design: A review. *Journal of the European Ceramic Society, 42*(1), 1–17.
12. Rafieipour, H., Setoodeh, A., & Lau, A. K. (2021). Mechanical and electromagnetic behavior of fabricated hybrid composite sandwich radome with a new optimized frequency-selective surface. *Composite Structures, 273*(114), 256.
13. Nazari, F., Taherkhani, M., Mokhatari, M., Aliakbarian, H., & Shekoofa, O. (2020). Efficient design methodology for sandwich radome panels: A C-band design example. *IET Science, Measurement & Technology, 14*(7), 808–816.
14. Shavit, R. (2018). *Radome electromagnetic theory and design*. John Wiley & Sons.
15. Editorial Team. (2024, April 7). What are air supported radomes? Retrieved from https://www.everythingrf.com/community/what-are-air-supported-radomes
16. Engineering Pilot. (2023, August). The radome: More than an "aerodynamic housing". Retrieved from https://www.engineeringpilot.com/post/the-radome-more-than-an-aerodynamic-housing
17. Communications and Power Industries (CPI). (n.d.). Ground-based radomes—ESSCO: Metal space frame radomes. Retrieved from https://www.cpii.com/product.cfm/12/184/889
18. Crone, G. A., Rudge, A. W., & Taylor, G. N. (1981). Design and performance of airborne radomes: A review. *IEE Proceedings F (Communications, Radar and Signal Processing), 128*(7), 451–464.
19. Harrington, R. F. (1968). *Field computation by moment methods*. Macmillan.
20. Jakobus, U. (2000). Comparison of different techniques for the treatment of lossy dielectric/magnetic bodies whithin the method of moments formulation. *AEU International Journal of Electronics and Communications, 54*(3), 163–173.
21. Houle, J. E., & Sullivan, D. M. (2020). *Electromagnetic simulation using the FDTD method with Python* (3rd ed.). Wiley-IEEE Press.
22. Silvester, P. P., & Ferrari, R. L. (1996). *Finite elements for electrical engineers* (3rd ed.). Cambridge University Press.
23. Delgado, C., Garcia, E., & Catedra, F. (2021). Fast hybrid computational technique for the analysis of radome structures using dual domain decomposition. *Electronics, 10*(18), 2196.
24. Meng, H., & Dou, W.-B. (2009). A hybrid method for the analysis of radome-enclosed horn antenna. *Progress in Electromagnetics Research, 90*, 219–233.
25. Moreno, J., Fernandez, M., Somolinos, A., Lozano, L., Catedra, F., & Gonzalez, I. (2013). Analysis and design of antenna radomes. In *IEEE International Conference on Microwaves, Communications, Antennas and Electronic Systems (COMCAS)* (pp. 1–5). IEEE.
26. Nair, R. U., & Jha, R. M. (2014). Electromagnetic design and performance analysis of airborne radomes: Trends and perspectives. *IEEE Antennas and Propagation Magazine, 56*(4), 276–298.
27. US Army Developmental Test Command. (2008). *MIL-STD-810G, Department of Defense Test Method Standard. Environmental engineering considerations and laboratory tests*. Aberdeen Proving Grounds.
28. US Air Force Life Cycle Management Center. (2015). *MIL-STD-461G, Department of Defense Interface Standard. Requirements for the control of electromagnetic interference characteristics of subsystems and equipment*. Wright-Patterson AFB.
29. Midya Chowdhury, A., Yadav, R., Chaudhary, V., & Panwar, R. (2021). Muhammad Talha Aamir, Muhammad Ali Nasir, Zafar Iqbal, Haris Ali Khan, Zhaib Muneer, Multi-disciplinary optimization of hybrid composite radomes for enhanced performance, Results in Engineering, Volume 20, 2023.
30. Midya Chowdhury, A., Yadav, R., Chaudhary, V., & Panwar, R. (2021). Broadband electromagnetic performance analysis of radome structures realized using hybrid equilibrium optimization strategy. In Intelligent Computing and Communication Systems (pp. 87-95). Singapore: Springer Singapore.
31. Dan, Y., & Yang, J. (2023). Electromagnetic performance analysis of inhomogeneous radome walls considering temperature and ablation. Aerospace, 10(11), 927.

32. Deng, J., & Zhou, G. (2017). A multi-disciplinary optimization design method for airborne radome. In *AIAA Modeling and Simulation Technologies Conference* (p. 4336).
33. Kolak, F., & Markina-Khusid, A. (2014). US Patent No. US8803733B2.

Aleksandra Markina-Khusid leads the Systems and Mission Analysis Department at MITRE Corporation. Her organization specializes in analytical and quantitative Systems Engineering and Mission Engineering (ME), including Systems of Systems (SoS) engineering, trade space analysis, and decision support under uncertainty, leveraging modern Digital Engineering (DE) approaches. Dr. Markina-Khusid began her career in computational electromagnetics, where she applied advanced modeling techniques to design and analysis of communications systems. She later pivoted to the engineering of complex systems, focusing on applications that address critical challenges for the Department of Defense and civilian US Government agencies. Dr. Markina-Khusid holds a BS degree in Physics, MS and PhD degrees in Electrical Engineering, and an MS in Engineering and Management, all from the Massachusetts Institute of Technology.

The Journey from Item Disassembly to Nuclear Power Generation and on to University Professor

Anne Lucietto

1 Introduction

In the 1960s, the landscape for girls and women in STEM (Science, Technology, Engineering, and Mathematics) was marked by significant gender disparities and challenges. Research has highlighted various factors contributing to the underrepresentation of women in math-intensive STEM fields during this period. These factors include cognitive ability, occupational interests, lifestyle values, field-specific ability beliefs, and gender-related stereotypes and biases [1]. Negative stereotypes about women's math abilities were prevalent and transmitted to girls by parents and teachers, impacting their attitudes and interest in STEM fields [2].

Studies have shown that exposure to exclusionary messages from peers within science classrooms could harm girls' intent to pursue STEM fields during adolescence [3]. Additionally, the intersectional experiences of Black women and girls in STEM education have been examined, shedding light on the unique challenges faced by this demographic group [4]. The importance of gender diversity in STEM fields has been linked to individuals' self-concept in mathematics, emphasizing the significance of representation in tertiary science education and careers [5]. Efforts to address the gender gap in STEM have included initiatives to support the attraction, retention, and progression of girls and women in STEM studies and careers [6].

Female role models have been identified as effective in both recruitment and retention phases in STEM, highlighting the importance of representation [7]. Strategies such as STEAM (Science, Technology, Engineering, Art, and Mathematics) education have been proposed to increase girls' engagement with STEM by encouraging interdisciplinarity, creativity, and innovation [8]. In conclusion, the 1960s marked a period of gender disparities and challenges for girls and

A. Lucietto (✉)
Purdue University, West Lafayette, IN, USA
e-mail: lucietto@purdue.edu

K. J. Bunker (ed.), *Women in Electrical Engineering*, Women in Engineering and Science, https://doi.org/10.1007/978-3-032-00133-7_3

women in STEM fields. Addressing stereotypes, biases, and lack of representation is crucial in promoting gender equity and diversity in STEM disciplines. Efforts to provide support, opportunities, and role models for girls and women are essential in fostering their interest and participation in STEM fields.

2 Parental Support

Parental support plays a crucial role in the development and well-being of children and adolescents across various stages of life, including childhood, adolescence, and college years. This section explores the multifaceted nature of parental support through examples drawn from research and interviews, mainly focusing on the insights provided by Ledo Lucietto in a Story Corp interview. Parental support significantly influences children's physical activity levels and overall health in childhood. A meta-analysis conducted by Yao and Rhodes highlights that parental encouragement and involvement are critical predictors of children's physical activity (PA) behaviors, suggesting that parents who actively promote and participate in physical activities with their children can foster healthier lifestyles [9]. Similarly, Liu et al. found that parents who accompany their children to sports and provide financial support for activities directly enhance their children's engagement in physical activities [10]. This aligns with the findings of Henriksen et al., who emphasize the importance of parental role modeling and awareness regarding children's PA [11]. These studies collectively underline the necessity of active parental involvement in promoting healthy behaviors from an early age.

As children transition into adolescence, the dynamics of parental support evolve but remain equally significant. Research indicates that various forms of parental support, such as encouragement and involvement, continue to play a vital role in adolescents' physical activity levels [12]. For instance, a study by Eime et al. illustrates that family support mediates the relationship between socioeconomic status and sports club membership among adolescent girls, indicating that parental engagement can mitigate barriers to participation in physical activities [13].

The insights from Ledo Lucietto's interview reveal that parental support during adolescence often includes emotional backing and guidance, which are essential for navigating the complexities of teenage life [14]. Transitioning to college represents another critical phase where parental support can significantly impact students' adjustment and success. Research by Lee and Fletcher indicates that first-generation college students often perceive lower levels of parental support compared to their continuing-generation peers, which can hinder their adjustment to college life [15]. This is echoed in the findings of Benito-Gomez et al., who argue that as students learn to navigate the college environment, they rely on their parents for emotional and decision-making support, which can enhance their overall college experience [16]. Moreover, the study by Hale and Buzas emphasizes the importance of structured parental involvement during the first year of college, suggesting that programs designed to educate parents about the college transition can empower them to provide more effective support [17]. The qualitative insights from Ledo Lucietto's Story Corp interview further illustrate the profound impact of parental support during the college years [14]. Lucietto describes how his father's encouragement and guidance were instrumental in helping him navigate the challenges of higher education, emphasizing the emotional and practical support that parents can offer during this transformative period. This narrative aligns with the broader literature, which suggests that parental support not only aids in academic success but also fosters resilience and independence among college students. In conclusion, parental support is a dynamic and essential component of child and adolescent development. From promoting physical activity in childhood to providing emotional and practical support during adolescence and college, the role of parents is pivotal in shaping positive outcomes. The evidence from various studies, coupled with personal narratives like that of Ledo and Elaine Lucietto, underscores the importance of fostering strong parent–child relationships throughout different life stages.

3 Life Story

Anne Lucietto's journey, from exploring her imagination in her early years to working in nuclear power generation and eventually becoming a university professor, is a remarkable story of perseverance, adaptability, and intellectual curiosity. Born and raised in a modest household, Lucietto's early experiences instilled a strong work ethic and a passion for learning. Her parents were supportive of their children's aspirations and encouraged them to pursue their dreams. Regardless of the circumstances, they supported her and her sibling as they navigated through life. The following shares bits throughout her life.

3.1 The Early Years

From a very early age, in this case about 4 years old, Anne wanted to help. One day, her father was making a grandfather clock and had been sanding the wood body while Anne watched. He left for a few minutes, and she decided to help by taking the sandpaper to the unfinished wood, unfortunately sanding against the grain. Needless to say, he was very unhappy when he returned to the project. In his usual manner, he asked what Anne had done. She explained, and he showed her the proper way to sand the wood. Despite his efforts, he was unable to completely fix the "little mess" she had made. Every time she looks at that clock, she is reminded of this lifelong lesson (Fig. 1).

When Lucietto was 6 years old, she was eager to attend her aunt's PhD graduation from the University of Chicago. She often spent time with this aunt and her grandparents, watching her aunt work hard as a doctoral student in the 1960s. Lucietto was often amazed at what could be done with a typewriter, as her aunt was

Fig. 1 Lucietto Family—Luigi (grandfather), little sister, Ledo (father), Lena (aunt), Anne, Onorina (grandmother), Lambert (uncle), and the clock

one of the first to have one of those fancy electric ones. Due to these observations, "Smith Corona" became part of Lucietto's vocabulary at a young age.

Lucietto agreed to attend the graduation on the condition that she remains quiet throughout the ceremony. As her aunt and other graduates walked down the aisle, Lucietto managed to stay quiet until she saw one of her favorite local children's TV personalities. Overcome with joy, she accidentally shrieked but quickly understood the event's significance and remained quiet for the rest of the ceremony. The experience left a lasting impression on her, as she remembered the beautiful gowns and the grandeur of the occasion.

When Lucietto was about 5 years old, the family created a costume for her, and everyone had a hand in it. They made sure Anne helped out, too. Lucietto was a pumpkin. Her dad and uncle made a frame for the pumpkin out of the wire they used at their shop, and Anne helped with the measuring and bending of the wire. Her mom made an orange pumpkin top and fastened it to the frame, helping with the construction. There was a debate on the best way to make a frame and stem as a hat, and they all worked on that together. Anne found long green socks and brown school shoes (Fig. 2).

Fig. 2 Lucietto as a child in the pumpkin costume made by the entire family

Lucietto went off to school in this costume and won the top prize! When the family wanted her to compete in the costume contest at her grandparents' Riis Park Field House in Chicago, IL, she won there, too! It was a family effort that was fun and resulted in quite a bit of recognition. They also taught her to integrate bending wire, sewing, and other skills.

She was known for tying her wagon to her bicycle and heading to the town dump. She'd heard many stories of her father and his brother scraping to make a few extra cents in their youth, so going to the town dump was perfectly acceptable to her. She was fortunate, as, during that time, people had been using and stopped using aluminum pots and pans for fear of developing dementia in their later years [18]. This was the boon she was looking for. She quickly investigated and learned various techniques to identify different metals easily and then began collecting them to take to the recyclers. That came to an end when she brought home a screen door. Her aunt quickly phoned her mom to tell her she had ridden by their house with quite a load. When she got home, mom was calm, no doubt thanks to the warning called in by her aunt. Her father showed her how to remove the steel parts, and they took it to the scrappers. Then he told her she had to pay for the gas to make these trips. The financial gain was so reduced that she decided to work on other ways to create and save money. Her mother later confirmed they were pleased that she made that decision herself.

Lucietto's family often took her to museums and activities promoting learning. Her father frequently suggested going to the zoo, with Brookfield Zoo and Lincoln Park Zoo in the Chicago area available to them. However, Lucietto always preferred going to the Museum of Science and Industry. One time, during grade school, their class had a field trip to the Oriental Institute at the University of Chicago. Lucietto was disappointed with this visit because she wanted to learn about the technical aspects of the excavation of King Tut's tomb [19], which wasn't available there.

When she got back home, she told her family that the Museum of Science and Industry was better, and she went to the library to learn more about King Tut. She felt that the field trip wasn't complete enough for her.

In those days, not all families had a TV. When her family decided to get one, she decided to take it apart. They were displeased with her, as it was an expensive purchase and their first TV. However, under pressure, she was able to put it back together just as quickly as she had taken it apart. She was rather organized and had taken it apart, laying the pieces in order of their removal. So reassembly wasn't as difficult as one might imagine. The only issue was that a flat round plastic piece with ridges on the edge never made it back in. After her father passed away many years later, she found the missing piece in his catch-all tray on top of his dresser.

She often took things apart or checked to see how things were connected—far more carefully each time. She had learned from the TV debacle. She later learned that they knew what she was doing. Her family did not chastise her for her activities fueled by inquisitiveness and the quest to learn. Rather, they supported her, even though that may have resulted in teeth gnashing!

When Lucietto started Junior High School, she already had a talent for disassembling things, understanding how they worked, and was more skilled in these areas than most of her peers. However, she wasn't interested in taking a Home Economics course, which was a requirement for Junior High completion. Instead, she wanted to learn about soldering, welding, metalworking, creating shop drawings, and working with wood. In those days, Home Economics and Shop Courses were segregated by gender [20]. Lucietto's father spoke with the school administration, and they agreed that she could attend the Shop class as long as she excelled and didn't cause any disruptions. Lucietto was thrilled with the arrangement and thoroughly enjoyed creating and building new things. She consistently outperformed her classmates and ranked among the top students in the class.

Junior High School ended with her winning geography competitions, participating in one-act plays as the lighting manager, winning awards in that, singing in the choir, and playing the piano for various performances throughout the years.

3.2 High School

High school was challenging for Lucietto. She went to a private high school where traditional gender roles were widely accepted. Throughout her 4 years there, she faced constant comments from faculty insinuating that, as a girl, there were limitations to what she could achieve. However, she knew this was false and she stood up against it.

Initially, she felt held back as it was hard to do the things she wanted in a strict environment. However, she decided to join various groups and participate in activities she might not have otherwise considered. It was fun that she took the train to school and got to meet many different and new people. She became a member of the choir, track team, and clubs focused on science and math. By the end of her freshman year, she was given the opportunity to compete in the State of Illinois Soloist Competition. She asked her parents for permission to attend, and they agreed. The competition was a one-day trip. Although she was uncertain about the competition, she chose to perform a dirge—"Hear Me! Ye Wind and Waves" by Handel. To her surprise, she ended up winning first place in the state—as a freshman or first-year student.

When she reached her second year of high school, she was involved in so many activities that she decided to quit track and focus on academic and musical pursuits. She excelled in science competitions and dedicated herself to improving her science projects each year. By the end of her junior year, she emerged victorious in the State Science Competition held at the Land Grant University in Illinois. Her main interests were in STEM and music, and she found joy and success in pursuing both.

In her final year, she struggled with being called a "truck driver" due to her way of speaking and interacting with others. This was mostly because of her experience working in shops and fighting against gender bias prevalent at the time. In defiance, she occasionally drove the family truck to school. She also focused on excelling in the ACT and SAT exams for college entry. Within a 6-h drive from her home, she applied to engineering schools and music programs and was accepted to all of them. Although her family and friends were thrilled for her to pursue engineering, she chose to attend a school for music and biology to eventually attend medical school, which was a complete departure from what seemed like a natural fit for Lucietto.

3.3 Undergraduate Years

At 18, Lucietto was a typical rebellious teenager, determined to carve her own path despite the expectations placed upon her. Everyone around her insisted she should pursue engineering, given her inquisitive nature, and aptitude for math and science. However, Lucietto was not ready to conform to these expectations without exploring her own interests first.

She enrolled in college with an open mind, initially applying to the engineering program to appease her family. Yet, her heart wasn't in it and she was responding to the constant comments by her family and friends. Instead, she found herself drawn to the arts and sciences, dabbling in music and pre-med courses. The allure of singing and the fascination with the human body captivated her, and she spent her first 2 years immersed in these fields.

During this time, Lucietto experienced the highs and lows of discovering her true passions. She enjoyed the creative freedom of music and the intellectual challenge of pre-med, but something was missing. She began to realize that while these fields were intriguing, they didn't fully satisfy her curiosity or align with her strengths. This really wasn't suited for her life course.

As she reflected on her journey, Lucietto recognized the value of her engineering background. The problem-solving skills, logical thinking, and innovative mindset she had developed were unique to engineering. She saw how these skills could be applied to real-world challenges and how they resonated with her natural abilities.

By the end of her second year, as Lucietto was encouraged by her father to complete college on time so he could focus on getting her sister through, she then made the decision to return to engineering. This time, it was her choice, driven by a genuine understanding of where her talents and interests truly lay. The experience of exploring different fields had given her a deeper appreciation for engineering, and she was ready to embrace it wholeheartedly.

Driven by a desire to further her education and pursue her interests in science and technology, Lucietto enrolled in college with the goal of studying engineering. However, she faced an unexpected obstacle; the engineering program was reluctant to accept a Liberal Arts student. Determined not to let this setback derail her ambitions, Lucietto devised a plan.

Every semester, she approached the dean as a Physics major, leveraging her advisor's support (one in Physics and one in Engineering) to take as many engineering courses as possible. She excelled in these classes, demonstrating her capability and passion for the field. Despite her successful performance, the dean remained hesitant to officially transfer her into the engineering program.

Lucietto's persistence never wavered. She continued her end of semester meetings with the dean, presenting her case and showcasing her academic achievements. Her advisors stood by her side, advocating for her transition into engineering. Finally, just before it was too late, the dean granted his approval.

The relief was palpable. Lucietto had overcome a significant hurdle, and her determination had paid off. She was now officially an engineering student, ready to fully immerse herself in the field she had fought so hard to join. This experience not only solidified her commitment to engineering but also taught her the value of perseverance and self-advocacy.

3.4 Career

Lucietto's journey was fueled by profound internal motivation. Her innate curiosity and passion for science and technology were the primary forces driving her relentless pursuit of knowledge. This intrinsic motivation was evident in her unwavering dedication to her studies and her determination to excel in a challenging field. For Lucietto, the internal drive was not merely about personal achievement; it was about making meaningful contributions to the world through her work.

After completing her undergraduate degree, Lucietto embarked on a career in the nuclear industry, where she applied her knowledge and expertise to tackle real-world challenges. Her work in the nuclear power generation plant (Fig. 3) allowed her to significantly contribute to the field, from working on large heating, ventilation, and air conditioning systems, to the installation of a chiller system dedicated to cooling the area around the reactors at the site. She was involved with the group

Fig. 3 LaSalle Nuclear Power Station, Marseilles, IL

focused on environmental activities such as the cooling systems utilizing fluid, air, and gas systems throughout the plant.

Anne Lucietto's career is a remarkable journey through various engineering and management roles, each marked by significant achievements and contributions. Her professional path began at Commonwealth Edison (ComEd) in Chicago, Illinois, where she spent a decade from 1985 to 1995. During this time, Anne served as a Project Manager and Principal Mechanical Engineer in the Materials Engineering Group. She was responsible for revising, specifying, and evaluating testing for a wide range of components, including metallic, elastomeric, plastics, petroleum, and chemical materials. Her role involved performing engineering analyses and consulting on diverse mechanical, electrical, and chemical issues.

One of her notable achievements was working with a team that reverse-engineered components used in power generation stations. This led to the development of a corporate shelf-life program that ensured the optimal usage period for all components, thereby reducing waste and improving reliability. During this time, she also got her ham radio license with her husband, who joined her mother and father as amateur radio operators!

Anne's responsibilities expanded when she became a Project Manager and Principal Engineer in Corporate Power Generation. In this role, she oversaw corporate inventory control, reliability issues, procurement activities, and the centralization of chemical purchases. She also managed the power generation receiving facility, ensuring efficient operations and adherence to corporate standards. Her expertise in outage management and cost control was evident as she developed solutions for corporate work schedule issues and trained generation plant personnel in scheduling techniques.

Despite her success in the industry, Lucietto remained dedicated to her pursuit of knowledge and continued her education by earning an advanced degree in business, focusing on finance and operations management. She was motivated to learn how to read financial statements and understand the discussions taking place among accountants and higher-ups in her organization. Her academic achievements opened the door to new opportunities, including research projects, teaching positions, and collaborations with leading experts in the field.

In 1996, Anne transitioned to IBIDEN Circuits of America in Elgin, Illinois, where she played a crucial role as a Project Manager and Facilities & Maintenance Engineering Supervisor. She was instrumental in building out a new manufacturing facility and establishing all related functions, ensuring that the plant was fully operational and met all necessary requirements. Later that year, she joined Driv-Lok, Inc. in Sycamore, Illinois, where she served until 2001. As a Senior Manufacturing Engineer, Applications Engineer, and Cost Estimating Supervisor, Anne oversaw technical interfaces, production engineering, and cost estimating functions. She supervised the estimating department, developed a revised cost estimating system that saved the company an and provided reverse engineering functions. Her ability to streamline processes and improve efficiency was a testament to her technical acumen and leadership skills.

Anne's subsequent significant tenure was at Fermi National Accelerator Laboratory in Batavia, Illinois, from 2001 to 2008. As a Senior Project Manager in Mechanical Systems Engineering and Facilities Engineering Services, she managed facility and maintenance functions, led subgroups on safety and environmental issues, and consulted on mechanical, electrical, and chemical issues. She managed specialized contracts and training programs, including a Reliability Program and a Crane/Hoisting program. Anne's leadership in these areas improved operational efficiency and enhanced safety and compliance standards. She successfully led multiple construction projects from inception through implementation, demonstrating her ability to manage complex projects and deliver results.

In 2008, Anne joined Caterpillar, Inc. in Aurora, Illinois, as a Senior Project Manager and Manufacturing Engineering Supervisor. She provided project management direction to manufacturing engineering, facilities/construction engineering, and layout planning. Anne coordinated capital expenditures for equipment construction projects, managed equipment installation projects, and led teams in updating corporate specifications to meet current standards. Her management of equipment installation projects, including demolition, restoration, and equipment moves for a facility spanning over 120,000 ft^2, was completed successfully under her leadership, ensuring minimal disruption to operations and optimal use of space and resources.

From 2009 to 2011, Anne worked at International Titanium Powder (Cristal Global US) in Ottawa, Illinois, as a Senior Project Manager, Plant Construction Manager, and Maintenance Manager. She managed all construction and maintenance activities, performed plant management tasks related to facility startup, and developed the Maintenance Department and Site Warehouse. Anne's strategic planning and day-to-day management of construction activities were pivotal in the successful installation and startup of the facility. Her ability to oversee complex projects and manage multiple responsibilities simultaneously highlighted her exceptional organizational and leadership skills.

3.5 *Professional Society Involvement*

During these years, Anne Lucietto was deeply involved in the Society of Women Engineers (SWE), where she held several key positions, including serving as the society treasurer. Her commitment to SWE extended beyond financial oversight; she also chaired various committees at the national level and served as president of the Chicago Regional Section. Lucietto's involvement in SWE was marked by her dedication to promoting gender equity in engineering and supporting the professional development of women engineers.

In addition to her work with SWE, Lucietto was actively engaged in industry-specific engineering societies. She held progressively responsible positions in mechanical, electrical, civil/construction, chemical, and environmental engineering. Her extensive experience in these fields contributed to her recognition as a Senior Member of the IEEE. This recognition, along with being named a Fellow in the Society of Women Engineers, reflects her significant contributions to the engineering profession and her commitment to advancing technology for the benefit of humanity.

3.6 *The Transition*

Anne will not make changes or pursue something new unless she has considered all of the consequences. Her academic journey is a testament to her dedication to both engineering and education. She began her higher education at Marquette University in Milwaukee, Wisconsin, where she earned a Bachelor of Science in Mechanical Engineering (BSME). Her studies focused on design, materials, and thermodynamics, laying a strong foundation for her future career. Anne then pursued a Master of Business Administration (MBA) at Lewis University in Romeoville, Illinois, where she specialized in finance and operations management. This combination of technical and business education equipped her with a unique skill set that would prove invaluable in her professional endeavors.

Lucietto's transition from industry to academia marked a pivotal moment in her career. She truly enjoyed working in the field, learning new things, and doing something different every day. However, the difficulty she continued to encounter in nearly every position that she held was the new graduates' inability to land on their feet when they graduated and were hired by either Anne or her company. So, she finally made the move to become a university professor.

Anne's academic pursuits culminated in a PhD in General Engineering with a focus on Engineering Education from Purdue University in West Lafayette, Indiana. Her dissertation, titled "The Role of Academic Ability in Choice of Major and Persistence in STEM Fields," was supervised by Dr. Matthew Ohland. This research explored the factors influencing students' decisions to pursue and persist in STEM fields, providing insights that would inform her future teaching and research.

3.7 Educator and Research Experience

Anne's teaching career is equally distinguished. She has been a faculty member at Purdue University since 2014, starting as a Visiting Assistant Professor before becoming an Assistant Professor and later an Associate Professor. At Purdue, she has taught a variety of courses, including Leadership and Organizational Change, Applied Fluid Mechanics, Power Plants and Energy Conversion, and Heat and Power. She has also mentored senior capstone projects, guiding students through the culmination of their engineering education. Anne's commitment to teaching extends beyond Purdue; she has been an adjunct instructor at Oakton Community College, Waubonsee Community College, and McHenry County College. At these institutions, she has developed and managed courses in statics, dynamics, strength of materials, algebra, engineering drawing, and various math and business courses. Her ability to teach both in-person, hybrid, and online has allowed her to reach a diverse student population.

In addition to her teaching, Anne has been actively involved in research at Purdue University. She has served as a Co-Principal Investigator (Co-PI) on a project funded by the Indiana Office of Energy Development, studying the impact of Small Modular Reactors (SMRs) on Indiana's energy portfolio. This research aims to diversify and modernize the state's energy sources, providing reliable and sustainable energy while preparing a workforce skilled in nuclear technology. Anne is also the Principal Investigator (PI) on a National Science Foundation (NSF) funded project titled "Collaborative Research: Gateway or Gatekeeper: Understanding Why Black Students Choose Engineering Technology or Engineering, and the Implications of this Choice." This study seeks to strengthen the partnership between academia and industry by using the experiences of Black students to recommend changes in how engineering programs are marketed, recruited, and taught to all students.

Anne's research interests stem from her dissertation on the relationship of SAT scores and persistence in STEM fields. She has continued to investigate the skills and identities of engineering technology students, focusing on issues such as math anxiety and reliance on intuition. Her work supports graduate students in projects related to renewable energy, mechanical design, and the use of infographics in education. Anne's research has practical implications, aiming to improve educational outcomes and support community involvement.

Throughout her career, Anne has also contributed to the development of industrial courses for various employers. She has created and taught courses in drawing and sketching, geometric tolerancing, statistics, physics, project scheduling, project management, and software applications. Her ability to translate complex technical concepts into accessible learning experiences has made her a valuable educator in both academic and industrial settings.

Anne Lucietto's journey from her early career in power generation to her current role as a respected university professor is a testament to her passion for engineering, her dedication to education, and her unwavering commitment to making a difference in the world. In this chapter, we explore the pivotal moments and transformative experiences that have shaped Lucietto's remarkable career trajectory.

As a university professor, she found fulfillment in sharing her passion for engineering and all that she had done with the next generation of students. Lucietto's dynamic teaching style, coupled with her practical insights from years in the field, inspired and empowered her students to explore new ideas and push the boundaries of scientific inquiry.

Throughout her journey, Lucietto remained guided by her curiosity and determination to make a difference in the world. Whether dismantling televisions, optimizing systems related to nuclear reactors, or mentoring aspiring engineers, she approached each challenge with enthusiasm and a commitment to excellence.

4 Summary

Anne's journey from television disassembly to nuclear power generation and on to university professorship is a testament to the transformative power of education, hard work, and a relentless pursuit of one's passions.

4.1 Struggles

Her story inspires all who dare to dream big and defy expectations in pursuit of their goals. Lucietto's path to success was not without its challenges. As a child, she often felt different from her peers. Her interests in dismantling televisions and exploring the mechanics of various devices set her apart. This sense of being different was further amplified when she was the only female student in her shop class. Despite the isolation and occasional skepticism from others, Lucietto's passion for engineering of all types never wavered.

4.2 Childhood Challenges

From a young age, Lucietto was fascinated by how things worked. She would spend hours taking apart old televisions and radios, much to the bewilderment of her friends and neighbors. While other children were playing outside, Lucietto was more interested in the inner workings of electronics. This curiosity sometimes made her feel like an outsider, but it also fueled her determination to learn and understand more. She recalls one summer when she spent weeks trying to fix an old radio she found in the attic. Despite numerous setbacks and failed attempts, she finally got it working. This small victory was a significant confidence booster and a testament to her perseverance.

4.3 Family Support

Throughout these struggles, Lucietto was supported by her family, who played a crucial role in her journey. Her parents, both from immigrant families (three grandparents were immigrants, one born to immigrant parents), understood the value of hard work and education. They encouraged her to pursue her dreams and provided the emotional and practical support she needed to overcome the challenges she faced. Her father, in particular, was a constant source of inspiration. He would often share stories of his own struggles and triumphs, reminding Lucietto that perseverance and dedication were key to success. Her mother, too, was a pillar of strength, always ready with words of encouragement and a listening ear. Lucietto recalls the regular dinners and family discussions about issues, new things, and whatever they wanted to discuss. Her family always had dinner together.

Growing up, family surrounded Anne. In the same small town, she could access her maternal great-aunt and cousins by bicycle. She would go see them and get assigned a job pulling weeds in her massive garden, after which she was then treated with great home cooking. Other times, she would walk over to her paternal uncle's home, try new foods like lasagna, and talk to her uncle. The support and wholistic environment this presented to Lucietto provided much of the needed support she got throughout her growing up years.

Anne's father and uncle (this same one mentioned) had their own business. Anne loved working there and learning new things that could be made from wire, how to do accounting, customer service, and just being with them at their shop. Yes, she was the only female—but that did not matter; everyone was treated the same.

4.4 Driven by Success

Lucietto's drive for success was also fueled by her family's immigrant background. The pursuit of the American dream was a powerful motivator, instilling in her a strong work ethic and a determination to excel. She recalls her grandparents working multiple jobs to provide for the family and create opportunities for their children. This background provided her with a unique perspective and a deep appreciation for the opportunities available to her, further driving her to achieve her goals. Lucietto often reflects on her parents' sacrifices and the values they instilled in her. Their resilience and determination became a source of strength for her, especially during challenging times. She remembers her father telling her, "Our family came here for a better life, and you have the chance to make it even better. Don't let anything hold you back." An example of this can be heard on Story Corp—https://storycorps.org/stories/ledo-lucietto-and-his-daughter-anne/.

Despite her struggles, Lucietto's internal motivation and the support from her family helped her persevere. Her journey is a testament to the power of resilience and determination and the importance of a strong support system in overcoming challenges and achieving success. Lucietto's story is not just about personal triumph but also the collective effort of a family striving for a better future. Her experiences highlight the importance of community, support, and the relentless pursuit of one's dreams.

4.5 Wrapping It Up

Lucietto's transition from industry to academia marked a pivotal moment in her career. As a university professor, she found great fulfillment in sharing her passion for engineering with her students. Her engaging teaching methods, combined with her practical experience in the field, inspired and motivated her students to innovate and challenge established scientific norms. Lucietto's journey emphasizes the significant impact of education, dedication, and the relentless pursuit of one's dreams.

References

1. Wang, M. T., & Degol, J. L. (2016). Gender gap in science, technology, engineering, and mathematics (STEM): Current knowledge, implications for practice, policy, and future directions. *Educational Psychology Review, 29*(1), 119–140.
2. Shapiro, J. R., & Williams, A. M. (2011). The role of stereotype threats in undermining girls' and women's performance and interest in STEM fields. *Sex Roles, 66*(3–4), 175–183.
3. Riegle-Crumb, C., & Morton, K. (2017). Gendered expectations: Examining how peers shape female students' intent to pursue STEM fields. *Frontiers in Psychology, 8*. https://doi.org/10.3389/fpsyg.2017.00329
4. Ireland, D. T., Freeman, K. E., Winston-Proctor, C. E., DeLaine, K., Lowe, S. M., & Woodson, K. M. (2018). (Un)Hidden figures: A synthesis of research examining the intersectional experiences of black women and girls in STEM education. *Review of Research in Education, 42*(1), 226–254.
5. Niepel, C., Stadler, M., & Greiff, S. (2019). Seeing is believing: Gender diversity in STEM is related to mathematics self-concept. *Journal of Educational Psychology, 111*(6), 1119–1130.
6. McKinnon, M. (2020). The absence of evidence of the effectiveness of Australian gender equity in STEM initiatives. *Australian Journal of Social Issues, 57*(1), 202–214.
7. Adams, E. C., Oduor, P., Wahome, A. M., & Tondapu, G. (2022). Reflections on two years teaching earth science at the women in science (Wisci) STEAM camp. *Journal of Women and Minorities in Science and Engineering, 28*(1), 23–39.
8. Ng, W., & Fergusson, J. (2020). Engaging high school girls in interdisciplinary STEAM. *Science Education International, 31*(3), 283–294.
9. Yao, C. A., & Rhodes, R. E. (2015). Parental correlates in child and adolescent physical activity: A meta-analysis. *International Journal of Behavioral Nutrition and Physical Activity, 12*, 1–38.
10. Liu, Y., Zhang, Y., Chen, S., Zhang, J., Guo, Z., & Chen, P. (2017). Associations between parental support for physical activity and moderate-to-vigorous physical activity among Chinese school children: A cross-sectional study. *Journal of Sport and Health Science, 6*(4), 410–415.

11. Henriksen, D., Mishra, P., & Fisser, P. (2016). Infusing creativity and technology in 21st century education: A systemic view for change. *Journal of Educational Technology & Society, 19*(3), 27–37.
12. Chen, P., Wang, D., Shen, H., Yu, L., Gao, Q., Mao, L., Jiang, F., Luo, Y., Xie, M., & Zhang, Y. (2020). Physical activity and health in Chinese children and adolescents: expert consensus statement (2020). *British Journal of Sports Medicine, 54*(22), 1321–1331.
13. Eime, R. M., Harvey, J. T., Sawyer, N. A., Craike, M. J., Symons, C. M., Polman, R. C., & Payne, W. R. (2013). Understanding the contexts of adolescent female participation in sport and physical activity. *Research Quarterly for Exercise and Sport, 84*(2), 157–166.
14. Lucietto, A. M., & Lucietto, L. B. (2008). You followed my footsteps and I thought it was great. *Story Corps*, NPR.
15. Lee, G. Y., & Fletcher, A. C. (2023). Emotional detachment moderates associations between students' perceptions of parental support and first-year college adjustment: Generational status differences. *Journal of College Student Retention: Research, Theory & Practice, 24*(4), 1002–1027.
16. Benito-Gomez, M., & Flores Rojas, K. (2020). Designing and implementing parenting interventions with Latino immigrant families: Challenges and strategies. *Families in Society, 101*(4), 528–538.
17. Hale, J., & Buzas, C. (2012). The parent first-year experience on a community college campus. *Journal of College Orientation, Transition, and Retention, 20*(1). https://doi.org/10.24926/jcotr.v20i1.2823
18. Álzate, K. G., Correa, D. A., Ortega, J. J. U., Berrio, L. F., & Cotta, L. A. B. (2023). Alzheimer's disease and its association with dietary aluminum: A review. *Tecnura: Tecnología y Cultura Afirmando el Conocimiento, 27*(77), 5.
19. Hindley, M., Lawler, A., & Lifson, A. (2015). King Tut: A classic blockbuster museum exhibition that began as a diplomatic gesture. *Humanities, 36*(5). (https://www.neh.gov/humanities/back-issues/vol36/issue5)
20. Bailey, B. L. (2004). *America in the seventies*. University Press of Kansas.

Anne Lucietto, PhD, is an engineer with vast experience in managing diverse workgroups across different environments and organizational levels, including facilities, construction, manufacturing, design, process, maintenance, and job shop. She has expertise in various engineering disciplines in manufacturing, campus/educational/research, and highly regulated environments. Anne is proficient in providing leadership and technical support in collaborative settings. She is currently a professor of engineering technology and works with students, helping them to successfully traverse their educational goals and move into their career of choice.

Microgrids and Mindsets: Bridging Energy Systems and Personal Resilience

Kaitlyn J. Bunker

1 Context: What is Resilience?

Resilience is broadly defined as the capacity to withstand or recover quickly from difficulties. The concept is relevant in many aspects of our lives, and this chapter will explore two specific applications of resilience: in electricity systems and in individual people. The United States Department of Energy defines energy resilience as "the ability of the grid, buildings, and communities to withstand and rapidly recover from power outages and continue operating with electricity, heating, cooling, ventilation, and other energy-dependent services" [1]. The American Psychological Association defines resilience as "the process and outcome of successfully adapting to difficult or challenging life experiences, especially through mental, emotional, and behavioral flexibility and adjustment to external and internal demands" [2]. While the opportunity to build increased resilience in our electricity system infrastructure and in our personal lives may require different approaches, there are interesting commonalities to explore that may support growth and awareness in both aspects.

2 Electricity System Resilience

One area where resilience is especially crucial is in our electricity system infrastructure. Electricity is the backbone of modern society, powering our daily lives and our economies. When there is a disruption in the availability of electricity, livelihoods

K. J. Bunker (✉)
Rocky Mountain Institute, Boulder, CO, USA
e-mail: kbunker@rmi.org

K. J. Bunker (ed.), *Women in Electrical Engineering*, Women in Engineering and Science, https://doi.org/10.1007/978-3-032-00133-7_4

and lives can be impacted [3]. Most electricity systems are designed to provide reliable electricity service, meaning that whenever electricity is needed, for example, whenever you flip on a light switch, electricity is available. It is important to note that there are areas of the world (including a significant population of 750 million people) where this is not the case; electricity is either not available at all, or is available unreliably for only certain unpredictable hours of the day [4].

While many electricity systems around the world do provide reliability, most are not designed to provide resilience. Meaning, most electricity systems are not able to bounce back quickly after an external shock that causes a power outage, in order to resume electricity services quickly and ensure that people have the electricity that they need. There are vulnerabilities in various components of the electricity system today, for example, depending on how electricity is being generated, if there is physical damage to the generation infrastructure or an inability to get the necessary fuel to the location of the generation infrastructure, it won't be possible to quickly restore the generation of electricity. Once electricity is generated, it needs to be moved to people who need it, typically through overhead transmission and distribution lines. If there is physical damage to those lines, people similarly won't be able to access electricity, even if it is possible to generate it.

2.1 Threats to Electricity Systems Globally

Around the world, there are different types of threats that might impact the electricity system and drive a need for resilient infrastructure that can bounce back quickly. For example in the United States, there have been wildfires that can cause damage to electricity infrastructure [5], or in some cases the power has been intentionally turned off in order to avoid fueling wildfires even further [6]. There have also been severe winter storms that cause unexpected and extended power outages, for example, in Texas in 2021 [7].

In other parts of the world, different extreme weather events can threaten electricity system infrastructure. For example in the Caribbean region, hurricanes with increasing intensity and occurrence have devastated island infrastructure, including the electricity system. In 2017, Hurricanes Irma and Maria occurred within a few weeks of one another, and caused devastation across several islands in the region. In Puerto Rico, it took 11 months to fully restore power to all communities after Hurricane Maria [8]. Figure 1 shows an example of damage in the British Virgin Islands following these same two storms in 2017.

Wildfires, winter storms, hurricanes, droughts, and other severe weather events are impacting electricity systems around the world today. The centralized nature typical of electricity systems leave them vulnerable to physical threats; however, there are opportunities to improve the resilience of our electricity systems and enhance our ability to restore power quickly in the event of a power outage, ensuring that people have access to the critical electricity that they need.

Fig. 1 Photo taken by author when working in the British Virgin Islands soon after Hurricane Irma in 2017; damage to the electricity system and vehicles is visible. (Photo credit: author)

2.2 *Opportunities to Improve Electricity System Resilience*

Given the crucial nature of electricity in our lives and the vulnerability of many electricity systems today to external shocks such as natural disasters, there is an opportunity to improve the resilience of our electricity systems. One option to address the vulnerability of overhead lines is to bury them underground. When this part of the electricity system is buried, it is no longer subject to high winds and other severe weather, although flooding and lightning strikes to exposed parts of the electricity system can still affect underground infrastructure. A main challenge with an undergrounding approach is the cost, which can of course vary greatly due to the terrain; one estimate in California is a cost of US$2 million per mile of buried power line [9].

Another opportunity is hardening existing or new infrastructure, meaning that it is designed and built to withstand relevant external shocks. This could include aspects of the design phase such as selecting the location for new electricity infrastructure with consideration for past hazards and risk going forward and ensuring that the installation of any equipment follows industry standards for weather risks that are relevant locally. Substations, for example, are an essential part of electricity systems, where transmission lines connect with distribution lines to transport electricity to communities; while there may be some upfront cost to hardening substation infrastructure, it can help maintain electricity service (or allow it to be restored quickly) [10]. Hardening of generation resources can also be impactful; for example, the experience of several islands during recent hurricanes informed the Solar Under Storm series of reports that highlight design and installation best practices and standards to ensure each solar photovoltaic (PV) project is as resilient as possible [11].

Another opportunity that is becoming more common is to build microgrids. Microgrids will be described further in the next section and offer an opportunity for the electricity system to become more distributed, diverse, and segmentable, which can reduce vulnerability to external shocks and increase system resilience.

3 Microgrids for Resilience

Many electricity systems today are centralized in nature, and they were built that way for good reason. Cost, efficiency, available space, etc. led to electricity infrastructure built with large generation plants, and long overhead transmission and distribution lines to bring electricity to people who use it. However, as described above, this leads to vulnerability in a world with increasing natural disasters. A microgrid is defined as "a group of interconnected loads and distributed energy resources that acts as a single controllable entity with respect to the grid" [12]. Microgrids can operate in connection with the grid, or can disconnect and operate independently as an electrical "island." Microgrids provide an opportunity for increased electricity system resilience through a more distributed architecture, through using a diverse set of generation resources, and through introducing an option to segment the electricity grid.

3.1 Decentralizing to Improve System Resilience

Microgrids are part of a more decentralized electricity system architecture, which can enhance resilience to external shocks. They can be deployed strategically to utilize the infrastructure currently in place and to focus on locations that can enhance operation of the electricity system during both normal and emergency times. For example, microgrids can be sited at critical facilities—those that serve the needs of communities both during a typical day and during an emergency such as an extended electricity outage. Examples include:

- Schools or community centers in the Caribbean that double as hurricane shelters
- Pharmacies and grocery stores that serve as a community hub during a disaster
- Telecommunications and water treatment facilities that maintain critical services and
- Hospitals or other medical facilities

When energy resources are located at these types of critical facilities, they can provide multiple benefits. During normal operation of the electricity grid, these energy resources can operate in connection with the grid to provide additional generation resources and help to balance the operation of the grid. During an extended grid outage, such as during a severe weather event, these energy resources can separate from the grid to maintain electricity to the critical facility forming a microgrid.

Many microgrids of this nature exist today as backup diesel generators that can turn on during a grid outage to keep a specific building operating. Utilizing renewable energy options for critical facility microgrids brings additional benefits, as they use local rather than imported resources and can provide value by continuing to operate outside of disaster events. Regardless of the type of electricity generation resource used, moving toward a more distributed system (compared to the common centralized system design used historically) can increase resilience.

3.2 Diversifying Sources to Improve System Resilience

Similar to a centralized system creating vulnerability because of the location of resources, relying on one single energy resource can also leave an electricity system vulnerable if there is an issue with that resource, in particular if the resource isn't locally available. Eliminating the single point of failure by diversifying the energy mix can improve the resilience of the electricity system.

Each location that might consider building a microgrid will have a unique set of local resources that make the most sense to utilize, together with the existing electricity grid infrastructure. Solar PV is often a top option as it is a proven technology used globally, and is very modular—one panel can be installed on a rooftop, or a large array can be installed on the ground, with many options for sizes and configurations in between. When paired with battery energy storage, solar PV brings an opportunity for communities to utilize their own local resources to meet their energy needs while increasing their resilience. Solar PV is of course only one option of many, including diesel or gas generators, wind, etc. Different resources often pair well with each other, with their different availability and costs to operate. Diversifying to utilize multiple resources rather than relying on any one single resource can strengthen the energy system overall.

3.3 Segmenting the Electricity Grid to Improve System Resilience

Microgrids can also increase resilience by introducing the option to segment the electricity grid. In a traditional electricity system, any issue in any part of the system (generation, transmission, distribution) means that a large portion of the system may be unable to operate. With microgrids that utilize a more diverse set of generation resources in more distributed locations throughout the system, there is a new option to operate only a portion, or a segment of the grid. This could look like restoring the power first to the most critical infrastructure following a storm, by operating a segment of the grid that contains local generation resources. Then, as any damage to other parts of the system is repaired, additional segments could be energized until the wider group of interconnected microgrids is fully operational again.

3.4 Microgrid Examples

Microgrids are in operation around the world today, providing resilience as well as other benefits to communities. One long standing example in the United States is in Borrego Springs, California, a community that is connected to the larger electricity grid through a single transmission line, leaving it vulnerable to extended power outages. The Borrego Springs microgrid is owned and operated by the utility company San Diego Gas & Electric and includes several solar PV arrays, battery energy storage systems, and diesel generators [13]. There are also examples of microgrids on the eastern coast of the United States, for example, in Brooklyn where microgrids have been built following the devastation caused by Superstorm Sandy [14].

While small island power systems don't fit the definition of a microgrid (because there is no larger electrical system to connect with), the experience of small islands in enhancing the resilience of their electricity infrastructure is relevant to larger geographies. There are many examples of small islands around the world leading in this area; one such project is in Ragged Island, a small island that previously required shipments of diesel fuel coming in regularly from Nassau, the capital of The Bahamas. Approximately 70 people live on Ragged Island, and their electricity system was destroyed in 2019 as part of the devastation caused during Hurricane Dorian. The island is served by the local utility company, Bahamas Power & Light (BPL) who decided after the storm to build a new system utilizing solar PV and battery energy storage, with diesel backup [15]. Ragged Island now uses 93% renewable energy, and BPL only needs to ship fuel every few months, much less frequently than with a full diesel-based system. Figure 2 shows a photo during the construction phase of the Ragged Island project.

4 Parallels to Personal Resilience

Separate from exploring opportunities to improve the resilience of electricity system infrastructure, there is a related concept of personal or individual resilience. Similar to having options for improving the resilience of electricity infrastructure (such as undergrounding, hardening, and microgrids as described in this chapter),

Fig. 2 Photo from Ragged Island during construction. (Photo credit Fidel Neverson, RMI)

humans have opportunities and options to improve our own resilience and be prepared before any type of external shock or challenge may occur.

To give one example as a working parent, I might identify an area of vulnerability in my life to be childcare. Perhaps I might have a plan for childcare most of the time, but am aware that it may be unreliable—if a teacher is absent with little notice, daycare may have to close for the day, for example. In order to boost my resilience in anticipation of potential challenges, I could do things like building a backup list of babysitters that I can call on, connecting with families in my neighborhood who could potentially share sitters with me, and working to increase my financial savings to prepare for the cost of emergency childcare needs. Similar to an electricity system, there are multiple options to improve resilience overall, and often multiple of these options can be pursued at the same time in order to increase the resilience of the system—whether that's energy infrastructure or my childcare situation.

In addition to opportunities to improve resilience in our daily lives, there has been significant research and work exploring the role of mindset in how we react to and bounce back from shocks that may occur in our lives. One framework for personal resilience to external shocks or challenges has been presented by Seligman [16]. He describes three common reactions to a struggle or a challenge in our lives, all starting with the letter P:

- Personalization: believing you are at fault for negative events
- Pervasiveness: thinking an event affects all aspects of life
- Permanence: feeling that the consequences of an event will last forever

People who are resilient have often developed a practice of addressing each of these Ps. Many or all of us would initially feel stressed or upset about a negative external shock in our personal world. However, it is possible to bounce back and demonstrate resilience through addressing one or more of personalization, pervasiveness, and permanence.

For example with personalization, we may go beyond taking accountability for our role in a situation and fail to see the role of external circumstances. We may assign blame or a sense of guilt to ourselves without reflecting on all of the factors involved. With pervasiveness, we may assume that a difficulty in one aspect of our life means that all areas of our lives are impacted. It can be easy to assume that a setback in one area impacts all areas of our life, when this isn't always the case. And with permanence, it can be hard to see beyond the current disruption or challenge. We may feel or believe that a crisis will last forever and miss an opportunity to begin working to improve the situation [17].

While some people are naturally able to work through challenging situations, all of us can learn and strengthen our abilities to become more resilient. Recognizing when thoughts of personalization, pervasiveness, and permanence sneak in is a first step, and can be followed by building a mindset of learning and action. In particular, positive emotion, engagement, relationships, meaning, and accomplishment (PERMA) can support resilience and growth [18].

4.1 Parallels Between Personal and Electricity System Resilience

There may be interesting parallels to draw between personal and infrastructure resilience. For example with pervasiveness as the concept that one event affects all aspects of life, this may be the case for a traditional, centralized electricity system where an issue in any part of the system affects the operation of the whole. However an external shock event doesn't have to affect the whole system if there is a microgrid supporting a more decentralized architecture. The distributed nature of electricity generation resources as used in microgrids can mean that a system can avoid pervasive challenges and avoid impacts across the full system.

With personalization as the concept of one person (often ourselves) being at fault for a large failure, there is a parallel to electricity systems that do rely on only one resource to generate electricity, or only one transmission line to get the electricity to where it is needed. A system that uses a more diverse set of resources can be more resilient as it's less likely for one component to contribute to a system-wide failure. Solar PV may be able to come back online quickly and serve the most critical loads, while overhead lines connecting other loads are rebuilt or fuel is obtained for other generation resources. There's potential to use this concept of diversity to strengthen our individual resilience as well; for example, recognizing that we have different components we can call upon both within ourselves and our communities, and that we are rarely the sole cause for negative or challenging events.

And with permanence as the concept of consequences lasting forever, an electricity system that has the option to segment its operation through the use of various microgrids can quickly move past this and ensure that at least the most critical infrastructure is able to get electricity relatively quickly after an outage. Perhaps the concept of segmenting can help to build individual resilience as well and overcome the perception of the permanence of negative events.

5 The Power of Women in Electrical Engineering

The field of electrical engineering provides many exciting technical areas of work, some of which are discussed in this chapter and throughout this book. Engineering is also not only about the technical or scientific aspects; there are also important components of creativity and working with people to understand their priorities and ways of thinking. By doing both, we can uncover relevant parallels and opportunities to connect our human experience with the engineering work we're exploring. We will need lots of people with diverse backgrounds and ways of thinking to continue advancing the field of electrical engineering and working to increase the resilience of both our electricity systems and ourselves as individuals.

In particular, women working in electrical engineering have made great contributions in technical topics, while also connecting with people and working together to find the best solutions. This has been one of my favorite parts of my career so far: being able to connect with and learn from people while applying my technical engineering skills to help creatively solve problems. In this chapter, I've presented one parallel that I've found fascinating, between personal and electricity system resilience. There are likely many other opportunities for different fields of expertise to learn from one another to continue advancing work of all kinds in support of our shared humanity.

References

1. United States Department of Energy. *Energy reliability and resilience*. Retrieved March 13, 2025, from https://www.energy.gov/eere/energy-reliability-and-resilience
2. American Psychological Association. *Resilience*. Retrieved March 13, 2025, from https://www.apa.org/topics/resilience
3. MET Group. *Importance of electricity: Powering modern life*. Retrieved March 13, 2025, from https://group.met.com/en/media/energy-insight/importance-of-electricity
4. International Energy Agency. *Access to electricity*. Retrieved March 13, 2025, from https://www.iea.org/reports/sdg7-data-and-projections/access-to-electricity
5. Union of Concerned Scientists. *Wildfires and power grid failures continue to fuel each other*. Retrieved March 13, 2025, from https://blog.ucs.org/paul-arbaje/wildfires-and-power-grid-failures-continue-to-fuel-each-other/
6. California Public Utilities Commission. *Public safety power shutoffs*. Retrieved March 13, 2025, from https://www.cpuc.ca.gov/psps/
7. University of Texas at Austin Energy Institute. *The timeline and events of the February 2021 Texas electric grid blackouts*. Retrieved March 13, 2025, from https://energy.utexas.edu/research/ercot-blackout-2021

8. Campbell, A. F. (2018, August 15). *It took 11 months to restore power to Puerto Rico after Hurricane Maria. A similar crisis could happen again*. Vox. Retrieved March 13, 2025, from https://www.vox.com/identities/2018/8/15/17692414/puerto-rico-power-electricity-restored-hurricane-maria
9. Institute for Energy Research. *PG&E to bury transmission lines at cost of $2 million per mile*. Retrieved March 18, 2025, from https://www.instituteforenergyresearch.org/the-grid/pge-to-bury-transmission-lines-at-cost-of-2-million-per-mile/
10. Burns & McDonnell. *A guide to hardening electrical substations for extreme weather events*. Retrieved March 18, 2025, from https://blog.burnsmcd.com/a-guide-to-hardening-electrical-substations-for-extreme-weather-events
11. RMI. *Solar under storm: Solar best practices for policymakers and two-part design guidebook to fuel resilient photovoltaic systems for small island developing states*. Retrieved March 18, 2025, from https://rmi.org/insight/solar-under-storm/
12. National Renewable Energy Laboratory. *Microgirds*. Retrieved March 19, 2025, from https://www.nrel.gov/grid/microgrids.html
13. San Diego Gas & Electric. *Borrego Springs microgrid*. Retrieved March 19, 2025, from https://www.sdge.com/sites/default/files/documents/2024-10/FINAL_S2470069_BorregoSprings_FS_ONLINE.pdf
14. The Guardian. *Brooklyn's social housing microgrid rewrites relationships with utility companies*. Retrieved March 19, 2025, from https://www.theguardian.com/sustainable-business/2017/aug/17/energy-brooklyn-social-housing-microgrid-rewrites-relationships-with-utility-companies
15. The Nassau Guardian. *$5 million solar project completed on Ragged Is*. Retrieved March 20, 2025, from https://www.thenassauguardian.com/home/5-million-solar-project-completed-on-ragged-is/article_4966358d-0365-59de-848b-ee9c95f19d89.html
16. Seligman, M. (2006). *Learned optimism: How to change your mind and your life*. Vintage Books a Division of Random House.
17. Diversity Australia. *Martin Seligman's framework for resilience 'The 3 Ps'—A framework for resilience*. Retrieved March 12, 2025, from https://www.diversityaustralia.com.au/martin-seligmans-framework-for-resilience-the-3-ps-a-framework-for-resilience/
18. Seligman, M. *Building resilience*. Harvard Business Review. Retrieved March 12, 2025, from https://hbr.org/2011/04/building-resilience

Kaitlyn J. Bunker, PhD, PE, is a director with RMI's Global South program, leading the Climate Finance Access Network (CFAN). In this role, she leads the team at RMI that serves as network coordinator for CFAN, which aims to unlock access to climate finance at scale in developing countries. Kaitlyn previously worked with RMI's Islands Energy Program, leading a diverse team partnering with islands in the Caribbean to support and accelerate their clean energy transitions.

Before joining RMI, Kaitlyn completed a PhD in electrical engineering from Michigan Technological University in Houghton, MI. Her dissertation research focused on microgrids and optimizing control strategies for distributed renewable resources. She stayed at Michigan Tech for graduate school following her undergraduate studies, which were also in electrical engineering.

With a mechanical engineer father and biochemist mother, Kaitlyn knew from a young age that she wanted to pursue a Science, Technology, Engineering, and Math (STEM) career. She received advice that electrical engineers use the most math out of all of the engineering disciplines, and as a math lover, she decided to pursue this option.

As a college student and later as a professional, Kaitlyn has been a long-time member of the Society of Women Engineers (SWE). Participating in SWE has allowed her to gain new leadership skills while contributing to an organization whose mission she is proud of. As a PhD student, Kaitlyn served as the Collegiate Director on SWE's Board of Directors, which opened up a new understanding of the importance of strategic thinking and collaboration. Kaitlyn received the SWE Distinguished New Engineer award in 2018, and served again on SWE's Board of Directors as Secretary in 2023–2025 and as a Director in 2025–2027.

Kaitlyn lives in Colorado with her husband and two daughters. They are slowly renovating their home to be as energy efficient as possible, with a goal to eventually be net-zero energy.

Launching a Diverse Career with an EE Degree

Tuyet-Hanh Schnell

1 The Spark

Electricity is a vital part of everyday life. From the very beginning, electrical pulses send information throughout our bodies driving our movements and processing external inputs from the outside world, so that it can run as proficiently as possible. Electricity powers devices and tools that people use daily often without giving it much thought. People rely upon electricity to help them perform hundreds of tasks for work and school. Nearly every industry has some reliance on electricity and the equipment and machines that use it.

I recall a family trip to the beach when we drove through a thunderstorm. I vividly remember seeing the lightning and wondering where it came from. Humanity's fascination with lightning and that display of raw energy have captivated people for centuries. They held lightning in awe and reverence. In the United States, many people will recall Benjamin Franklin's kite experiment, where he flew a kite with a key on its tail to prove his concepts of electricity [1].

My father, two aunts, and a second cousin worked at the Bendix Electric Power Division in Eatontown, New Jersey. My father was a technical writer, and I got to see some of the intricate drawings that he was incorporating into his technical manuals. My aunts, one a diligent clerk and the other a skilled field coil winder for electric generators, would delight me with stories of their time on the factory floor. My second cousin, a meticulous production tester, would often talk about the importance of precision in manufacturing.

But it was a single visit to the facility on Family Day that truly ignited a passion within me. As I stepped onto the production floor filled with huge generators and production material, I was intrigued by the number of people and the different skills

T.-H. Schnell (✉)
Lockheed Martin, Moorestown, NJ, USA
Rowan University, Glassboro, NJ, USA

K. J. Bunker (ed.), *Women in Electrical Engineering*, Women in Engineering and Science, https://doi.org/10.1007/978-3-032-00133-7_5

required to create these products. That day, something clicked, and a spark was lit—a spark that would stay with me for the rest of my life.

A pivotal summer program in Science Technology Engineering Math (STEM) at Brookdale Community College solidified my interests in electrical engineering. That summer, I spoke with electrical engineers at Fort Monmouth and Bell Laboratories and electrical engineering students at NJ Institute of Technology and Rutgers University. Although my uncle was a civil engineer, he had died before I was born so I did not know that the engineering field was an option for me. Therefore, this exposure not only broadened my understanding of the field but also introduced me to the Army Signal Corp Boys Scout Explorer Program at Fort Monmouth, which offered hands-on lessons in electricity to both boys and girls.

I was eager to be a part of this program, and, with my parents' permission, I joined. These weekly sessions provided a solid foundation in electronics, which I built upon by taking every circuits class available in high school. I was thrilled to design, etch, and fabricate my own circuit boards including creating a functional pig-light game and transistor radio—both of which remain in working condition to this day. Given my passion and experience, choosing to study electrical engineering at Stevens Institute of Technology, in Hoboken, New Jersey, was an easy decision for me to make.

2 Making the Connection

During my time at Stevens, I seized the opportunity to join the Co-operative Education program, despite it being in its inaugural year. I was drawn to the program's unique blend of course work and hands-on real-world experience, allowing me to apply engineering principles in diverse corporate settings while still in college. This exposure deepened my understanding of electrical engineering and renewed my passion for learning.

As a trailblazer in the inaugural graduating class of the Co-operative Education program, I had the unique opportunity to provide feedback and contribute to the program's evolution. By sharing my experiences and insights, I was able to help shape the program's future direction and suggest improvements, ultimately benefiting the students who followed in my footsteps.

I had three co-op assignments. The first was at Vista Research, Inc. which was a small company based in California. But I worked in a small office located at the Environmental Protection Agency's facility in Edison, New Jersey. I supported research of fiberglass and steel underground storage tanks [2]. In this role, I developed macros in spreadsheets to parse through collected data from controlled leak tests. I also wrote FORTRAN programs to plot the results of deformation testing on the tanks based on temperature and environmental conditions. I also created the safety documentation for the research site and presented the safety protocols to all the employees at the test facility. Upon completion of this first assignment, I received the framed certificate shown in Fig. 1.

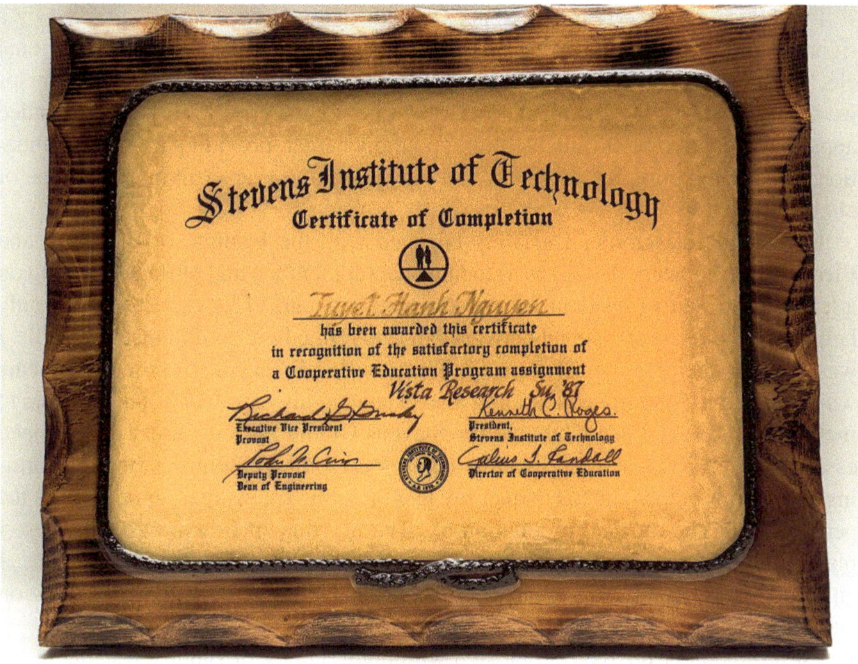

Fig. 1 Certificate of completing my first assignment

My second co-op assignment was with Georgia Tech Research Institute at Evans Area Fort Monmouth. I was a junior technician working on radar research to develop and prototype of a new chip to perform signal processing. I worked with engineers from Fort Monmouth to create schematic diagrams of the chip design and develop simulation software to test the chip performance and behaviors. I also had a chance to evaluate differences in working for the government as a civilian or working as a government contractor. I found that I enjoyed working for a government contractor better than working directly for the government.

My final co-op assignment was with General Electric Astrospace in East Windsor, New Jersey. I was a junior designer working on the Global Positioning Satellite II-R design for the U.S. Air Force. I was given my own project—the startup module of the Satellite Processing Unit (SPU) box, which handled and processed data collected by the satellite's sensors and instruments. I was able to successfully take the design from start through to Critical Design Review with the customer. It was quite eye opening for me to see the design process in a real-world setting. Being able to see the customer reaction to my design was also an amazing experience. Re-use is something that industry leverages to reduce the cost of production of new products. When the team working on the satellites for the American Telephone & Telegraph (AT&T) Company saw my design for the startup module, they decided to re-use it for their satellite. To this day, I am amazed that I was able to create something that was used in so many satellites while I was still in college.

In addition to my co-op experience, my education at Stevens was marked by a rigorous and interdisciplinary approach, with a curriculum that spanned traditional electrical engineering courses in circuits, physics, and calculus, as well as courses in mechanical, civil, chemical, and software engineering. This well-rounded approach proved to be invaluable throughout my career, providing a strong foundation for tackling complex problems and navigating the ever-evolving landscape of engineering.

Beyond the classroom, I actively engaged with the Institute of Electrical and Electronics Engineers (IEEE) as a student chapter member, and ultimately taking on a leadership role as president for two consecutive years. This experience not only helped me build a professional network but also refined my communication skills. I was also the recipient of scholarships from the IEEE for three consecutive years which helped to fund my education. I also expanded my extracurricular activities to include the Stevens Glee Club and the Stevens Drama Society, where I developed essential soft skills that are critical in team-based settings. Being in these performing arts groups helped me to conquer my fear of public speaking. Figure 2 shows me performing in the musical production of "The Mystery of Edwin Drood" where I played the character of Rosa Budd. By embracing these diverse opportunities, I cultivated a well-rounded set of skills that complemented my technical education and prepared me for a successful career.

I still remember the day I graduated with my Bachelor of Engineering degree in Electrical Engineering from Stevens Institute of Technology. I sang the national anthem to open the ceremony with my family and friends there to celebrate with me.

Fig. 2 Schnell performing a solo

Fig. 3 Schnell graduates from Stevens Institute of Technology

It was a moment of great pride and accomplishment, and I was filled with excitement and anticipation for the future. Little did I know at the time that the foundation I was laying would ultimately form the backbone of my career—like a backplane in electronic systems. Figure 3 shows me on graduation day.

In complex electronic systems, the backplane serves as a framework that seamlessly interconnects circuit boards and modules, allowing them to communicate and function in synchronization. Similarly, my education and experiences were creating a robust framework that would support my professional journey, providing a common thread that connected the disparate threads of my knowledge and expertise.

Just like a well-designed backplane that enables the efficient exchange of data and resources between various components, my educational foundation was enabling me to integrate and access my skills and knowledge across different roles and situations. This framework would prove to be a vital asset, allowing me to adapt, grow, and thrive in an ever-evolving professional environment.

A piece of advice from one of my earliest technical mentors has stayed with me throughout my career: "School teaches you how to learn. Mentors will teach you everything else." My academic foundation has provided a springboard for continuous learning and growth. In the workplace, I've been fortunate to have many mentors who have shared their expertise and knowledge, guiding me on the tools and skills necessary to excel in my engineering roles.

3 Starting as a Hardware Engineer

My first position after graduation was that of an Associate Member of Engineering Staff at General Electric Government Electronic Systems Division. A prime example of applying what I learned in school to my job was using the theoretical foundations I gained from Stevens to design through-hole surface-mounted digital modules for signal processors and programmable array logic chips. My coursework in digital circuit logic helped me create board and chip designs, while I also delved deeper into the application of differential equations in performing stress analysis. The research skills I honed during my senior design project enabled me to conduct thorough trade studies to select the components for my board designs.

Using my membership with the IEEE, I leveraged their vast library of technical information and connections with fellow electrical engineers. This affiliation has kept me informed about the latest innovations and technologies, allowing me to stay up-to-date and incorporate best practices into my designs and testing methodologies.

During one of my assignments, a senior engineer shared valuable insights into the inner workings of the engineering lab, particularly in the controlled environment of clean rooms. This guidance helped me navigate the intricacies of lab etiquette, communicate effectively with technicians, and adhere to strict lab protocols. Through mentorship and hands-on experience, I developed a comprehensive understanding of design principles to effectively isolate power and clocks across different layers, minimizing the impact of crosstalk on my designs. With newly found knowledge and confidence, I collaborated with electrical technicians to refine and perfect my chip and board designs before they entered production.

As my printed circuit boards arrived from the production line, I thoroughly tested them in the clean room environment to ensure they met project expectations. If any issues arose, I provided technical feedback and recommendations for improvement. Unfortunately, the clean room's shared usage among multiple projects often required me to work unconventional hours, including testing periods that began at 4 a.m. I was simultaneously pursuing a graduate degree at Drexel University. In those days, you had to be in the classroom. So going to class from 6 p.m. to 9 p.m. and then waking up early to test at 4 a.m. made for a very intense schedule and forced me to prioritize my time carefully.

My department wanted to introduce a new computer-based design layout tool for the engineers in the department. This tool would be used to design and place components as well as draw the connection layers of the circuit board. Since Stevens was one of the first colleges to require that each student has a personal computer, I was very comfortable with learning new software products. Therefore, my manager sent me to a weeklong class on the use of this new software tool to gain knowledge and then return as the subject matter expert (SME) on the tool for my co-workers.

With only a few months at my company, I went on my first business trip and interfaced with other engineers from different companies that were also using this tool. In my new SME role, I gained self-confidence and an increased sense of ownership. I utilized my social skills obtained from hours of serving the students at Stevens as a leader in numerous extracurricular activities on campus to interact with the other electrical engineers in my department.

While I was learning skills from mentors, I became a mentor to others in this new role. This role also made me more visible to different technical leaders in my department who noticed my abilities and gave me more responsibilities and opportunities to grow. These opportunities included working in the research and development of new signal processing technology and developing field programmable gate arrays (FPGAs). I also became a mentor of two new engineers, teaching them the skills that I had acquired. To this day, I still serve as a sounding board for one of these engineers.

Reflecting on my more than 30 years in engineering, I realize that my mentor's advice has been a constant theme. While school provided a solid foundation, it is the cumulative experience and knowledge gained from mentors, work experience, and life-long learning that have truly defined my career.

4 Taking on a New Discipline

When work in hardware became light after being in the department for 4 years, I needed to leverage my network, including members of the company choir (Moorestown Music Makers) that I joined, to find other opportunities within the company. My job search led to a position within the software engineering department. With software education already in my backplane of knowledge, I was comfortable learning to program in C and Ada 83. The manager that interviewed me decided that I would be a good fit for the position that he had open but also wanted me to try out the role for 6 months before agreeing to transfer to the new department. At the same time, GE sold my division to Martin Marietta. I was in the same position at the same desk, but I was given a sticker to put on my badge with the new company name to cover the old one.

The new role consisted of writing diagnostic software to determine the readiness and health status of the Aegis Combat System. Coincidentally, some of the hardware that would be tested was the same ones I was already familiar with having worked on those systems in the hardware department. With my electrical engineering background and my experience with the systems being tested, I was able to seamlessly adapt to my new work environment.

I was able to bridge the knowledge gap of the software team performing the work while learning how to write software in the new programing languages. My domain knowledge helped my new team maintain progress instead of waiting for answers from the systems engineers. After being in software engineering for 6 months, my new manager and I decided to make my transfer permanent. The manager responsible for this new work immediately saw my leadership potential and appointed me the Responsible Engineer for one of the software products and a team of two junior engineers to lead. It was during this time that Martin Marietta merged with Lockheed Corporation to become the Lockheed Martin Corporation.

As I embarked on my first leadership role as an engineer in the corporate world, I couldn't help but feel a mix of excitement and nerves. Although I had previously held leadership positions in college, this was a new chapter in my career, and I was eager to prove myself. I successfully led that team and created new documentation to help the entire software team.

With that success, I was given an even larger role, taking over the leadership of a program. I worked with the customer to define our scope and worked with systems engineers to create the associated requirements. When our work package was completed, I led the demonstration of the new project to the captain of the ship that will receive this new product and his commanding officers at the customer's test facility in Virginia. I did the demonstration of this new capability while I was 7 months pregnant with my first child. The captain and officers were thrilled with the capabilities of this new updated product. My efforts in this role were recognized by the Aegis customer community with the Aegis Excellence Award shown in Fig. 4.

My manager's vote of confidence in my abilities helped ease my concerns that led to my success, but it was a piece of advice from a trusted mentor that has had a lasting impact on my career journey from this moment onward. He counseled me to always look for my replacement—a suggestion that initially left me confused. However, as I reflected on his words, I began to understand the wisdom behind them. In essence, he was encouraging me to develop a deep understanding of my roles and responsibilities and to identify the skills required to excel in my position. As I mastered these skills, I was also to identify individuals on my team or among my peers who possessed some of these skills and showed potential to fill my role in the future.

Fig. 4 Aegis excellence award

By doing so, I was able to groom a suitable replacement, freeing myself to focus on high-priority tasks that leveraged my unique skills and expertise. This approach would also enable me to create a development plan for my successor, allowing them to grow and learn while I pursued new areas of interest. Moreover, it would help me articulate my contributions and added value to the organization, even as I transitioned out of the role.

My first use of this advice came as I was preparing to go out on my first maternity leave. By systematically cataloging all the work responsibilities that I had, I was able to provide my manager with a comprehensive list of tasks and the associated skillsets to perform those tasks. Because I gave my manager sufficient notice of my expected time out of the office, I was able to work with him to select a replacement for me. I also had time to train her and allow her to slowly take over certain critical tasks so that she could still ask me questions before I went out on my maternity leave. Although I had worked hard to get to this first delivery with my customer, my delivery of my first child was more important. Handing over that effort let me leave work with peace of mind and a clean slate to return to work after my leave of absence. This advice has proven to be a valuable guiding principle in my career, and one that I continue to draw upon to this day.

5 Addressing Change

Just as electrons in a dynamic system are perpetually in motion, reacting to the forces that surround them, our careers are also subject to change. External factors, such as industry standards and government regulations, can exert forces that redirect our focus. Working in the aerospace and defense industry, we are governed by expectations built into our requests for proposal. As the government seeks to regulate and demand the best from industry, it will specify standards that must be met.

One of those changes came in the late 1990s when the new government contracts required companies to perform software development at an assessment level equal to or higher than Level 4 (Managed Level) in the Software Engineering Institute (SEI) Capability Maturity Model (CMM) [3]. The model was intended to guide software developers in controlling their processes and focus on a limited set of improvements to those processes to increase product quality and maintenance capability.

Since I was the project manager of the Readiness Product on one of the projects that would undergo this SEI assessment, I needed to fully document our process and work with my team to ensure that our documentation and artifacts were in proper order and take action to obtain any missing data. Additionally, I served as the chairperson for the process review board to review and approve any process changes for the Readiness Product. This position helped me further hone my leadership skills as I often had to resolve differing and sometimes opposite opinions to formulate a suitable solution.

Since this assessment had never been conducted at my site before, it was a great undertaking to prepare for the assessment. I was interviewed by the assessment team during their site visit to make sure that my answers were consistent with the documented process and associated objective evidence. Once the assessment team completed their review, the Moorestown site was given an assessment of SEI Level 4. For my contribution in this new endeavor that laid the foundation for successful future assessments, I received an Evening of Excellence Award—the highest level award that the Moorestown site confers (Fig. 5).

6 Creating a New Path

As I grew in my leadership responsibilities both in team and budget size, I began to see a new path for my career—that of a functional leader. I began to prepare myself to be able to take on that new role. I spoke to my mentors about the skills needed to be a leader and took on roles that would give me that experience. I looked internally for job postings of software managers in my company. I also spoke to my mentors about my goal of becoming a first line leader.

Finally, an opportunity arose when my manager accepted a role in the program office leaving his position vacant. I immediately applied for the role because I had leadership experience and product domain knowledge. I was given an interview but did not receive an offer because one of my teammates who was more senior on the team also applied and got the offer to become the new manager.

Fig. 5 Outstanding achievement award

The feedback from the interviewers was that I was a strong candidate, but the other person had more overall experience. At that point, I was satisfied with the response and focused on getting my team to an important demonstration milestone. Because of this high-profile, time-intensive project, I did not actively look for another management opportunity.

However, since my mentors knew of my goal, one of them called me to let me know of a management position in another software department that he thought I would be a great fit to fill the role. Unbeknownst to me, he also spoke with the hiring manager to tell him that he should interview me because he thought I was the right person for the job.

Despite the demands of preparing for a critical production demonstration with a key customer, I seized the opportunity to apply for this leadership role and earned myself an interview that would prove to be a turning point in my career. As fate would have it, my unique blend of interpersonal and leadership skills perfectly aligned with the vision of this leader who was searching for the ideal candidate to nurture this team.

An engineer at heart, I approached this new opportunity by looking at the problem statement and then worked to determine how to address the issues. To me, leading and managing a group of engineers who are my direct reports are not that different from solving engineering design problems—the subject is shifted but the process remains the same. The first thing that I did was to speak with the manager who previously led this group to get an overall sense of the group, learn about the team's skillsets and identify who the key personnel were on the team. I then met with the stakeholders to understand the expectations of the team and where the team was or was not meeting expectations. Then I met with each individual team member to get an understanding of their current workload, their aspirations, and their unique needs. I also met with the teams' technical experts to get smart about the product features and capabilities. Armed with this information, I was able to development an action plan for me to acclimate with my new team and stakeholders and for my team to address the issues and concerns voiced by the stakeholders.

By using an engineering problem solving mindset I was able to discern the needs and wants from all the affected parties, I gained the knowledge to resolve the existing issues for the stakeholders and, at the same time, create a better environment for my team. The result was greater customer satisfaction and improved team morale. An unexpected outcome of this step was the development of a cross-training program for my team to increase their skillsets and ability to swarm issues when they arise. After spending 18 months in the role, I went into labor nearly a month ahead of my due date. Thus, I needed to take maternity leave unexpectedly early and had to quickly transition the organization to another manager. This situation was not ideal because integrated remote work and collaborative tools were not very mature in the early 2000. However, the new manager and I made it work through numerous phone calls, emails, and the use SharePoint folders.

7 Ready for Something New

After a 3-month leave of absence, I returned to work to take on the leadership of a small 8-person group performing internal research and development. This small team was a spin off from another group that had gotten too large for one manager to handle. Just like my experience with my previous role, I was coming into a product area that was unfamiliar to me. Anyone who has worked in research and development (R&D) already knows the uncertain nature of funding. Successful R&D projects require understanding the external market and product differentiators. Although I had participated in R&D in the past, this was my first time leading a team in this type of work. I once again used the scientific process and problem solving to acclimate to this new role.

This small, but mighty team had successfully developed a new way to develop command and control software and was able to patent that process. Interestingly, this group had three projects that were all lead by women engineers. As I settled into this new role, I found myself in uncharted waters of this matriarchy and loved every moment. Being in a male-dominated field, this new organization was a breath of fresh air for me.

One of the best ways for R&D to show value is to be used in a customer-funded program. As opportunity would have it, a proposal arrived and one of the products that our group developed was to be used for this new proposal. I was also selected to be the manager in charge of getting the cost estimates from all the engineering departments supporting this proposal. I was excited for this opportunity to work with all the engineering departments and further develop my network. I also worked closely with the proposal technical team to ensure that the design and engineering scope being proposed was properly captured in the cost estimates provided by their respective departments. Before the proposal was submitted to the customer, I presented our overall cost estimate to the vice president of engineering for his approval, sharing my assessment of the engineering costs along with the risks and their mitigation. With the vice president's approval, our proposal was sent to the customer who ultimately chose our submission. We had won new business for my company with a new customer.

I was given another opportunity to expand my network when my manager asked me to take his place as a member of the software sub-council to represent my work location. This software sub-council consisted of representative of software engineering from every company location in the United States. This group met quarterly at different company locations to discuss prevailing topics for software engineering, to update and approve changes to our overall software engineering policies and procedures, and to plan an annual software engineering conference for the entire corporation.

When I was at Stevens Institute of Technology, I had helped organize our annual IEEE day where engineers from local companies presented topics and demonstrated products to students. I was also a professional member of the Society of Women Engineers (SWE), the Society of Hispanic Professional Engineers (SHPE), and the

Society of Asian Scientists and Engineers (SASE), so I had attended several large conferences. Therefore, I was familiar with technical conferences and how they should be run. So when the software sub-council asked for volunteers to serve on the 2007 Agile Conference, I jumped at the chance to contribute. I served on the committee to review abstracts and curate topics for the conference and create the conference schedule. I also served as the moderator for one of the panel discussions on Agile adoption success within Lockheed Martin. Lastly, I organized the ice breaker event to start off the conference. The conference was a huge success, and many great takeaways and connections were made.

8 Taking the Logical Next Step

My next logical step was to advance in functional leadership and take on a leader of leaders role. It can be said that timing is everything, and for me, the timing was right. For the past several years, my manager had been providing me with roles and work experiences to prepare me for that next step. He even had me temporarily step into his leadership role when he was placed on a special assignment for several months. If luck is what happens when preparation meets opportunity, then I have been very lucky indeed in my career. As I was looking for this next assignment, a position as a senior manager of the systems engineering organization responsible for command-and-control (C2) products became available. I had worked with managers and engineers in this organization in the past to support several proposals, programs, and customer meetings. Therefore, I was quite familiar with their charter. I took the risk and applied for the position. After interviewing with human resources and a panel of other senior managers, I was offered the position. I excitedly acceptedly and sought to forge the next chapter in my career. This role gave me a large organization of more than 100 engineers with five managers. Shortly afterwards, my organization grew to include two additional groups: one at a location in Virginia and another at a location in the Minnesota. Figure 6 shows my organization.

9 Going Beyond the Work

The next several years were filled with successes, setbacks, and many lessons to learn. A major success was my being asked to serve as the lead for the women's employee resource group at my location. I had been leading the SWE group for a few years and was asked to take over this new role when the former leader left the company. I accepted the role and merged the two organizations to take full advantage of resources from each group called the Women's Impact Network (WIN). I also expanded the leadership board to provide more leadership roles as a training ground for future business leaders. I also began to interface and work with other leaders at other locations to share ideas and collaborate on events and activities. The

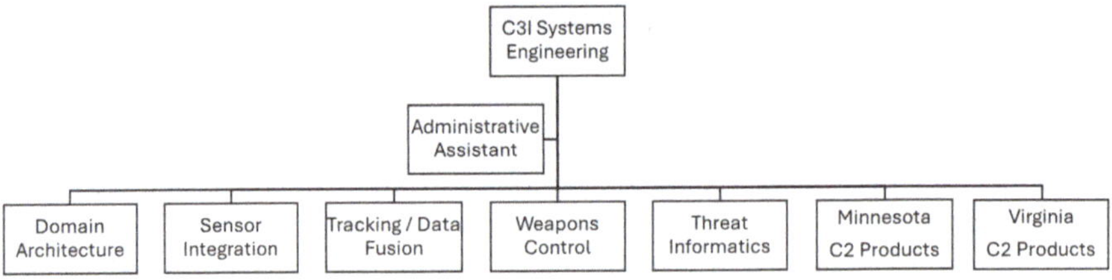

Fig. 6 Schnell's C31 organization

Fig. 7 Schnell spotlighted during Evening of Stars Awards Event, 2013

work that I did to empower women and foster synergy across work sites earned me a major award at my company, the Evening of Stars—Excellence in Diversity Award. Figure 7 shows my being spotlighted during the awards celebration. I also went on to lead WIN across twelve locations and served on the planning committee for several leadership conferences.

Another major success was the creation of the Desktop Test Suites for testing our C2 products without needing to utilize end-item testing assets. This testing platform that emulated the customer equipment allowed engineers to remotely log into the test suite and run integration testing and verification without added travel time to drive to the physical test locations where the real hardware resided. This suite also provided a way for engineers to create automated test scripts that could run overnight saving countless hours of physical labor. This testing capability provided a better work environment for engineers and cost savings to the programs.

Some setbacks that I experienced were the decline in work for my department. Despite having quality engineers and strong proposal submissions, we were not gaining as much new work as we needed to keep the staff fully employed. Therefore, over the course of 3 years, we had to exercise three reductions in force actions and the closure of the entire Minnesota work site. The morale in my department as well as other departments was at an all-time low.

At that time, there were two C2 systems engineering departments focused in different areas and products. With the loss of work, it became clear that both departments needed to combine under one leader. I had kept my technical skills current while in my leadership role by serving as a requirements expert on engineering review panels across my business area. I supported new proposals, program startups, and interim program reviews to evaluate the project's engineering performance health, identify risks and opportunities, and working with those teams to ensure project's overall performance success. I felt that I should step away from functional leadership and return to my technical roots once again.

But before I left my leadership role, I worked with the other senior manager to reorganize our two teams into one team. Using my Lean Six-Sigma Green Belt certification knowledge, I also worked with my counterpart to create a new organizational structure that would streamline project leadership to demonstrate to our customers that we understand how to operate as a lean organization. We also created a focused product leadership organization that would allow for enhanced architecture and more cross training across the product teams. These organizational changes were implemented across all products, not just C2, and remain in place to this day.

As much as going into leadership for the first time scared me, going back to a purely technical role was also unnerving. I supported several small projects as an individual contributor until there was an urgent need for systems engineering help at a site in Orlando, Florida. My manager offered them my assistance as a systems engineering and requirements expert. This 6-month assignment was exactly what I needed to ground my nerves. The project was the development of the ground-based F-15SA pilot and maintenance training systems for the Royal Saudi Air Force. The press release shown in Fig. 8 describes the scope of this contract.

These training systems were developed to teach pilots and weapon systems officers on aircraft avionics both in a desktop environment and in a dual-seat cockpit mockup. The training system also included a simulator that trains the crew on egress in an emergency. The final component of the training platform is the application of virtual training for the maintenance of the F-15SA without needing the physical aircraft. The design of the training facility was also part of the tasking for the team (Fig. 9).

I approached this new assignment with the systems engineering mindset to help me acclimate quickly with the new team in a location where I had little prior reputation to draw from. However, as a leader within the WIN, I knew people at every large site within my business area. Therefore, I reached out to the WIN site lead in Orlando to get an understanding of the work culture and overall infrastructure of the location. This way, I would not be arriving to this new group unprepared. I also reviewed as much documentation that was available for the project before my first trip to the Orlando facility. I also learned that the team was mostly software engineers, so I drew from my years of experience as a software engineer to communicate with that team and get an understanding of what they needed from systems engineering.

Lockheed Martin To Provide Virtual Training Technologies In Support Of Saudi Arabia's F-15SA Modernization Program

PR Newswire
ORLANDO, Fla.

ORLANDO, Fla., March 11, 2013 /PRNewswire/ -- Lockheed Martin (NYSE: LMT) received a $253 million contract agreement to begin work on F-15SA pilot and maintenance training systems for the Royal Saudi Air Force. The technologies will provide a comprehensive ground-based training environment for Saudi Arabia's F-15SA modernization program.

Pilots will complete air-to-air combat, air-to-surface missions, air combat maneuvers and tactical intercepts with 360-degree full mission trainers. As a first for F-15 training, the systems will feature a single dome over the dual-seat cockpit to enable crew coordination training. Lockheed Martin will also deliver egress, avionics and desktop trainers for procedure training by pilots.

"The complement of F-15SA training systems starts with desktop trainers and progressively increases in capability to full mission weapons systems trainers," said Jim Weitzel, vice president of training solutions for Lockheed Martin's Mission Systems and Training business. "This creates a cost-effective program since the appropriate level of technology is applied to meet the training objectives."

For maintainers, Lockheed Martin will provide virtual systems to enable training without the actual aircraft. The systems include basic maintenance, landing gear and arresting hook, armament, flight controls and jet fuel starter trainers.

Lockheed Martin will deliver the training systems by 2020. Work will be performed in Akron, Ohio, and Orlando, Fla. The contract is managed through a foreign military sales agreement by the U.S. Air Force Life Cycle Management Center at Wright-Patterson Air Force Base, Ohio.

Lockheed Martin has served as the Royal Saudi Air Force's F-15 training partner for more than three decades. The company also provides the service's weapons systems officer training program. In addition to delivering training systems to Saudi Arabia, Lockheed Martin's F-15 training technologies are currently in operation by the United States, Israel and Japan.

Headquartered in Bethesda, Md., Lockheed Martin is a global security and aerospace company that employs about 120,000 people worldwide and is principally engaged in the research, design, development, manufacture, integration and sustainment of advanced technology systems, products and services. The corporation's net sales for 2012 were $47.2 billion.

For additional information about Lockheed Martin, visit our website: http://www.lockheedmartin.com

SOURCE Lockheed Martin

Fig. 8 Press release for the F-15SA training contract [4]

Because the work included creating designs for the physical simulators as well as the network and physical layout of the training facility, I was able to pull from my knowledge of electrical engineering to review these designs and provide my feedback to the team. I also provided valuable inputs to the team regarding IEEE standards that the team needed to utilize to properly design the hardware system. At the end of my assignment, I had created a brand-new network Prime Item Development Specification (PIDS) and modified 11 other PIDS to give the software team the requirements that they needed to create the software simulations. Furthermore, these requirements were also testable within the timeframe and budget of the contract.

Being able to pull my knowledge of software, hardware, communications, and systems engineering confirmed to myself that I had not made a mistake returning to a technical role after 14 years in management. I was ready for the next phase of my career which focused on a technical path once again. With the completion of this systems engineering assignment, I was ready to tackle the next part of my career journey.

COMMERCIAL AEROSPACE

Lockheed Martin to support Saudi Arabia F-15SA modernization with virtual training technologies

ORLANDO, Fla., 13 March 2013. Lockheed Martin (NYSE:LMT) engineers will begin work on ground-based F-15SA pilot and maintenance training systems for the Royal Saudi Air Force, based on a $253 million contract under Saudi Arabia's F-15SA modernization program.

March 13, 2013

VIEW IMAGE GALLERY

ORLANDO, Fla., 13 March 2013. Lockheed Martin (NYSE:LMT) engineers will begin work on ground-based F-15SA pilot and maintenance training systems for the Royal Saudi Air Force, based on a $253 million contract under Saudi Arabia's F-15SA modernization program.

NEW

Army surveys industry for expertise in hyperspectral imaging sensors for unmanned aircraft and satellites

Software-development for design and development tools worst-case execution time analysis introduced by LDRA

Amazon launches first operational satellites for Project Kuiper from Cape Canaveral

MOST READ

United Airlines backs blended wing body startup JetZero

Space Force eyes overhauling 3 million lines of obsolescent software for next-generation nuclear-monitoring

Special Operations Command asks industry for uncrewed systems and data processing for special forces uses

SPONSORED

N-Type Coaxial Connectors in A Nutshell

Fig. 9 Excerpt from Military + Aerospace Electronics Magazine Article [5]

10 Leveraging the Past

The next role would also pull from another experience from my past, that of agile development. Aegis product development organizations were embracing agile development, and the Weapons Product Systems Engineering department was the first organization to roll out this new way of doing work. They were looking for people with agile development experience to help them properly adopt this endeavor. Because I had been in the software organization that utilized agile development techniques in the past and had served on the 2007 Agile Conference planning committee, I became the right choice to be a scrum master for this organization as it was adopting agile tenants for the first time. I joined the department as a Lead Member of Engineering Staff and served as the scrum master for their Scheduler product.

I was excited to start this new role and learn a different product area. Although I did not have any experience with weapons control software, I did have basic knowledge of how engagements were handled within Aegis. I was fortunate to have joined a product group that looked to me to help them adopt agile practices and were willing to teach me about their product and how it interfaces with other weapon products and the overall Aegis Weapon System.

At the same time, I was asked to support the roll out of agile development to the other product areas by becoming an agile instructor and coach. I wholeheartedly embraced that opportunity and gained two new mentors who were Lockheed Martin Fellows and experts in agile development. I was able to job shadow and co-teach with them until I was confident to teach on my own. For the next several years, I taught agile workshops every month to teams as they began to adopt agile methodologies. Additionally, I traveled to customer locations to provide briefings to them on agile practices as they were adjusting to our new business rhythm. These customers included high ranking officers from the U.S. Navy, the U.S. Coast Guard, and the Missile Defense Agency, along with civilian leaders and technical personnel.

My teaching then expanded to support my business area that encompassed 35,000 employees. I traveled to various locations in my business area to present the Program Initiation Workshops to teams and programs throughout the United States. In this role, I also became a mentor to other agile practitioners who wanted to become agile coaches. Now, my mentees shadowed me and had opportunities to co-teach with me. Through this process, I developed four new agile coaches for Lockheed Martin.

11 A New Opportunity

In 2018, I was given the opportunity to travel to South Korea to support the installation and check out of the latest software delivery to one of their Aegis destroyers. I was thrilled to travel to a new country and represent my department and its products with the customer. I spent hours in the Combat System Engineering Development Site (CSEDS) lab associated with this shipset to train for the event and worked with the other travelers to understand our roles and responsibilities for the trip. I also reviewed information about traveling to South Korea to understand its culture and how to interact with the South Korean people and the sailors and officers on the ship that I was supporting. On this assignment, over the course of 2 weeks, each product representative had to perform system testing and verify that the new software was installed properly and was in operational condition and ready to support the ship and its crew. We even went out to sea to fully test the capabilities of the radars and sensors on the ship. Foreign travel also gave me an opportunity to experience a different country and its culture. During my time off from work, I visited historic sites such as the Gyeongbokgung Palace in Seoul (Fig. 10).

Having been successful supporting a foreign military sales program in South Korea, an opportunity arose for me to support another foreign program. This time, it was developing capability for a brand-new class of ships for the Japan Self-Defense Forces. The lead for the project was inexperienced in leading such a large development effort, so I was asked to take over because of my vast experience in leading teams throughout my career. While I was hesitant to accept the leadership role again because I was enjoying the freedom of performing as an individual

Fig. 10 Schnell at Gyeongbokgung Palace, 2018

contributor, my manager convinced me that I was needed to ensure that the project succeeded.

Unlike leading from initial project start, taking over someone else's plans when the project was behind and over budget was a challenge. Again, I applied the systematic approach of working with my predecessor to understand the scope of work and his thought process for developing the execution plan. I then performed an assessment of the remaining work and created a return to green plan. After sharing my plans with the program leadership, I was given approval to execute that new plan. I was successful and the project portion for my product was back on track.

Over the next 6 years, I would take the Weapons Products from initial Aegis Light Off (first installation and integration of hardware and software on the ship) to completion of multiple live fire events at sea and sailing on these ships with the Japanese crew in support of those events (Fig. 11).

Along the way, I developed a great working relationship with the Japanese officers, sailors, shipbuilders, and technical consultants. I had a chance to see the beautiful country of Japan through the numerous trips there to support the customer. Figure 12 shows a photo I took of Mount Fuji from the Narukawa Art Museum. I also used the opportunity as a leader on the program to train and teach junior engineers both in the Weapons Product and in other product areas. As a result of the leadership and technical accomplishments that I obtained, I was promoted to Principal Member Engineering Staff, a title that less than 4% of engineers in my business area achieve.

Fig. 11 Schnell at shipboard Vertical Launching System, 2022

Fig. 12 Photo Schnell took of Mount Fuji at Hanoke's Lake Ashi, 2021

12 Continuous Learning Mindset

Wanting to continue to learn and grow in my technical knowledge, I applied for and was selected to lead the Weapons Product team on a demonstration program of a new missile. It was exciting for me to learn new data analysis tools using Matlab and to create some script updates in python, a language that I had not previously used. Because of my diverse background supporting ship installation and live firing events in the past, I was able to bring a lot of knowledge of expected behaviors of the system. I was able to explain why the system performed in the matter that it did and the reasons behind the product design. We successfully demonstrated this new missile capability in a live fire event at sea. We then performed analysis of the collected data and presented the results and recommendations to the customer.

Upon completion of this project, I was selected to be the Systems Engineering Integration & Test (SEIT) Lead for a distributed sensing internal research and development project. In this project, I was returning to my roots in electrical engineering and radar design. I also had the opportunity to learn from five LM Fellows who were also supporting this project. They encouraged me to apply to the LM Associate Fellows program because I had a lot of knowledge and experience.

After speaking with my manager and senior manager, I was asked to fill out my self-assessment against the LM Fellows set of criteria. I also worked with my LM Fellow mentor to properly document and highlight the accomplishments I had throughout my more than 30-year career. We also highlighted the numerous internal and external awards that I had received over the years.

At the same time, I was encouraged to also apply for Fellow Grade with the Society of Women Engineers (SWE) because of my technical body of work and the 20 years of service and leadership that I had with SWE. In October 2024, I received the SWE Fellow Grade Award at the annual conference in Chicago, Illinois. This award is SWE's Apex level or highest level of award that they confer each year. Figure 13 shows myself accepting the award.

In January 2025, I received the Lockheed Martin Associate Fellow title and attended the LM Fellows Conference. There, I met other LM Fellows to learn from and collaborate with while we heard company strategy to see how we can help achieve them. Figure 14 shows me at the LM Fellows Conference. Being an LM Associate Fellow, it was important for me to be a driving force in advancing crucial programs for the company. One of these initiatives was called 1LMX, which is Lockheed Martin's mission-driven business and digital transformation program. In this program, I worked to transform the end-to-end business processes and tools with a fully integrated digital thread throughout the product lifecycle. I joined to 1LMX Model Based Engineering Tools Adoption Team to help create the systems engineering development environment for Aegis programs, which became an early adopter of the tools suite. In doing so, I helped to lay the foundation and formulate lessons learned for the next program to adopt these new tools.

At this point in my career, I also felt that I needed new ways to grow the next generation of engineers. Thus, I applied to be an adjunct professor at the Electrical

Fig. 13 Schnell receiving her SWE Fellow Award, 2024

Fig. 14 Schnell becomes Lockheed Martin Fellow, 2025

and Computer Engineering department at Rowan University. I interviewed with the head of the department and was accepted I was asked to teach a graduate level Fundamentals of Systems Engineering course in the spring semester of 2025. Remembering what it felt like when I was a student, I was happy to share my knowledge and experience to these students. Being able to pull from my career experiences to show my students how these principals can be applied to real-world activities help them to connect the material they are learning with their use in industry. As this course had students that worked full-time in engineering roles, I encouraged them to share their experiences with other students in the class during class discussions. I added a second course to my teaching course load in the fall semester of 2025. This new course pulled from my technical experience to teach War Gaming with Command, Control, Communications, Computers, Intelligence, Surveillance and Reconnaissance (C4ISR) at both the undergraduate and graduate levels. I loved the challenge of bringing my industry experience as well as my experience working with the U.S. Navy, the Missile Defense Agency, and the Department of Defense into the classroom to show students the relevance of what they were learning as it applies to solving real world problems.

13 Final Thoughts

My career has certainly been an amazing journey of self-discovery, following my dreams, and listening to advice of others. As a lovely highlight to my career, I shared my career journey as the opening Keynote Speaker to six hundred attendees at the Society of Women Engineers WE Local Cherry Hill Conference in Cherry Hill, New Jersey, on March 7, 2025 (Fig. 15). As I continue my career as a member of the LM Fellows community, I will continue to innovate new technologies and grow the talented engineers who will follow me. I will continue to reach new students at

Fig. 15 Schnell as Keynote Speaker at SWE WE Local Cherry Hill, 2025

Rowan University and give them a strong technical foundation on which to create the solutions and innovations we have not yet imagined.

I hope that my story has resonated with you. I hope that you can take away from it some small nugget of wisdom that can help you in your own career. Remember to bring others along with you as you grow and rise in your career. Looking back, I am happy to be able to see how I have influenced technology. But more than that, I can see the impact that I had on the trajectory of others' careers.

References

1. Viemeister, P. (1961). *The lightning book*. MIT Press.
2. https://www.sciencedirect.com/science/article/abs/pii/030438949185025I
3. Paulk, M. C., Curtis, B., Chrissis, M. B., & Weber, C. V. (1993). Capability maturity model, version 1.1. *IEEE Software, 10*(4), 18–27. https://doi.org/10.1109/52.219617. Retrieved from https://insights.sei.cmu.edu/documents/1092/1993_005_001_16211.pdf
4. https://news.lockheedmartin.com/2013-03-11-Lockheed-Martin-to-Provide-Virtual-Training-Technologies-in-Support-of-Saudi-Arabias-F-15SA-Modernization-Program
5. https://www.militaryaerospace.com/commercial-aerospace/article/14226819/lockheed-martin-to-support-saudi-arabia-f-15sa-modernization-with-virtual-training-technologies

Tuyet-Hanh Schnell is an Associate Fellow and Principal Member Engineering Staff. She leads the research and development efforts at Lockheed Martin Rotary and Mission Systems in Moorestown, New Jersey. She oversees the technical, cost, and schedule for the design, development, integration, and test of the software products across multiple product domains. In her more than 35-year career at Lockheed Martin, she held numerous technical leadership roles including Technical Director, Integrated Product Team Lead, Responsible Engineer, Scrum Master, and Subject Matter Expert. She is an adjunct professor in the Electrical and Computer Engineering Department at Rowan University. She became a Society of Women Engineers (SWE) Fellow in 2024. She is a recipient of the WE23 Prism Award, WE22 Advocating Women in Engineering Award, WE Local 2022 Engaged Advocate Award, and WE Local 2021 Legacy Award. She received a special recognition award at the Women of Color Conference in 2022, Society of Asian Scientists and Engineers (SASE) Career Achievement Award in 2024, and numerous Lockheed Martin internal awards throughout her career. She was honored by the U.S. Navy Technical Representative with an Aegis Excellence Award for Outstanding Performance in 1997. Her biography is featured in, "Asian American Women in Science: An Asian American History Book for Kids." Schnell holds a Bachelor of Engineering from Stevens Institute of Technology and a Master of Science in Electrical Engineering from Drexel University. Schnell is a certified Lean Six-Sigma Green Belt. She is a life member of SWE, and is an active member of SASE, Women in Manufacturing, and the National Management Association. Schnell serves on the South Jersey District Board of Directors for Junior Achievement of New Jersey. She served as President of the SWE Southern New Jersey Section from 2023 to 2025 and a member of the several SWE Committees, the Asian Connections Affinity Group, and the Late Career and Retiree Affinity Group. She can often be seen performing STEM outreach in her community where she records hundreds of hours annually. She sings in her company choir and cantors at her church. Schnell resides in New Jersey with her husband, who is also an engineer, and their two children.

Integrating Electric Vehicles Successfully into the U.S. Electric Grid

Jill S. Tietjen

1 Introduction

The expected large-scale penetration of electric vehicles (EVs) on the U.S. electric grid will have a variety of impacts on the operation of each of the components of utility systems—generation, transmission, and distribution. The nirvana of so-called vehicle grid integration (VGI) is showcased in a July 2024 U.S. Department of Energy report that envisions the many pieces and parts working together successfully. Much work still remains to be done to reach that nirvana including resolving a wide array of challenges. These challenges include those addressed in this chapter: (1) the many stakeholders that must work together to make VGI a success, (2) the impact of the timing of charging on the need for utility resources, (3) discharge of EVs during the day as an energy storage mechanism, (4) impacts on the distribution systems, (5) forecast of anticipated growth in EV charging needs over the coming years, and (6) alternative charging mechanisms.

2 Utility Assets

Three categories of assets enable electric utilities to provide the electricity demanded by their customers as it is needed, when it is needed. The asset categories are generation—power plants or other energy resources including distributed energy resources (DERs); transmission—high voltage wires and some connections to individual customers (generally large industrial customers); and distribution—lower voltage wires and equipment that provides connections to most customers.

J. S. Tietjen (✉)
Technically Speaking, Inc., Greenwood Village, CO, USA

K. J. Bunker (ed.), *Women in Electrical Engineering*, Women in Engineering and Science, https://doi.org/10.1007/978-3-032-00133-7_6

When EVs charge at someone's home or in the airport parking lot or at the parking lot of the mall or retail store, the charging occurs on the distribution system. Fleet charging for vehicles like school buses, truck fleets, or bus fleets will probably occur on the distribution system as well. Large-scale charging plazas as envisioned for the future when EVs constitute a significant portion of the vehicle fleet may occur on the transmission system.

3 Types of EV Chargers

The integration of electric vehicles into the grid critically relies on the technology used for charging those EVs. As currently configured, there are multiple levels of EV chargers—Level 1, Level 2, Level 3, Level 4, and the DC fast chargers. These are commonly abbreviated as L1, L2, L3, L4, and DCFC as shown in Table 1. Note the significant differences in the size of these chargers and the locations of use which determine whether they can connect to the distribution or transmission

Table 1 Electric vehicle charger characteristics and impacts [1, 2]

Types of charger	Size	Characteristics and impacts
L1	2.6 kW	Single-phase AC supply generally using a three-pin socket. Slowest of chargers available to end users. Can be plugged into a standard 120-V electrical outlet at a home. Provide 2–5 miles of range per hour of charging. Commonly used for overnight charging
L2	7.4–50 kW	A range of 7.4 kW with a single-phase ac supply or 11 kW with a three-phase AC supply (maximum for a home). A DC supply allows for up to 50 kW. These chargers require specialized equipment to be installed regardless of supply. Can be plugged into a 208- or 240-V electrical outlet, similar to those used for large household appliances like electric dryers and ovens. Provide approximately 30 miles of range per hour of charging. Commonly found at public charging stations and workplaces
L3	60–350 kW	A range of 60–100 kW for "rapid" and up to 350 kW for "ultra-rapid" chargers. These require a strong tie to the distribution substation or a connection for transmission service depending on the facility size and the total chargers in the location. These are generally public charging hubs or parking lots offering charging stations as an amenity for their users
L4	1 MW or greater	These "megachargers" are in research and development and are planned for the trucking and other heavy equipment industries. Any facility of size (10 or more stalls) would require significant ties to a dedicated distribution substation and will likely need transmission infrastructure updates
DCFC		Provide fastest charging speeds for EVs. Have a wide range of charging speeds. Can add as many as 200 miles in 15 min. Typically found along major highways

system. The resulting impacts on the distribution or transmission system of the affected electric utility system are not trivial with large-scale penetration of EVs and need to be addressed. L4 chargers are currently in the research and development phase.

4 Vehicle Grid Integration (VGI) Stakeholders

Successful VGI will require the efforts of many—women and men—across many fields of endeavor—including electrical engineers! The U.S. Department of Energy (DOE) identifies 13 different categories of stakeholders and provides information on the roles they will play in the VGI effort as shown in Table 2. The DOE emphasizes the need for collaboration among the stakeholders for VGI to be successful. An examination of the many roles and responsibilities leads one to project that VGI will take years to develop.

For one of these stakeholder groups—utilities and grid operators—the North American Electric Reliability Corporation (NERC) in January 2024 examined the state of industry readiness and concluded "there are significant gaps in both the electric utility and EV charging community's technical understanding, planning and modeling of EV charging characteristics." NERC indicated that actions had been taken to address the knowledge gaps but much more collaboration would be needed. NERC identified deficiencies in modeling, VGI programs, standardization, and studies [2]:

- **Modeling**: As it currently stands, there is only a single, generic electrical model to represent EV charging. More work is needed to ensure that the electric system planners and operators have the quality of models needed.
- **Vehicle-to-grid programs**: Operators have the need to integrate vehicle-to-grid programs and impacts into their operational planning assessments or real-time assessments such that the assessments reflect policymaker-approved equipment standards for their footprint. EV original equipment manufacturers (OEMs) should strive for greater standardization and integrate emerging technology into current operator systems, especially since EVs cross operator and/or interconnection boundaries.
- **Standardization**: Currently, EVs and their charging systems do not follow consistent control philosophies or performance. Simply put, two different EVs that use Level 2 chargers do not necessarily interact with the grid in the same way. This lack of standardization makes grid planning difficult. Efforts are under way within the electric utility industry to address this issue.
- **Transmission studies**: There are no robust transmission studies to capture the impact that EVs, and their charging systems have on common planning problems. Through beta studies that focus on EV impacts and improved study procedures, NERC anticipates that transmission providers will need to undertake recurring studies to accurately capture reliability risk for their areas [2].

Table 2 Stakeholder roles with respect to VGI [3]

Stakeholder	Role
Utilities and grid operators	Plan for electricity infrastructure upgrades to meet growing EV charging demand, enable vehicle connections, develop markets and programs to compensate participation, propose electricity rates, and collaboratively develop standards
Utility regulators	Implement legislative direction to meet individual state EV goals, determine utility compensation mechanisms, and assess utility infrastructure investment plans and proposed electricity rate tariffs. Regulators establish program requirements and set objectives
Ratepayer advocates, community groups, and intervenors	Intercede in regulatory proceedings to promote the interest of consumers to maintain affordability for all customers; highlight opportunities to use advanced technologies and highlight opportunities for policy action to drive equitable solutions
State and local governments	Set energy, transportation, and building goals or targets; develop state transportation electrification plans and implementation programs; establish local zoning codes; and determine community development decisions that affect the load demand utilities must plan to meet
Aggregators	Develop mechanisms for aggregating EV loads alongside other DERs to provide grid services in response to system constraints or to help mitigate peak demand, interact with customers, participate in standards setting processes, and develop solutions that appeal to different customers' preferences
Vehicle Original Equipment Manufacturers (OEMs)	Determine EV charging control capabilities, set default settings, and integrate protocols and standards that enable vehicle interoperability with chargers and the grid so that vehicles can provide grid services. They engage with customers to understand vehicle and charging preferences
Charger manufacturers	Develop charging equipment, its capabilities, and related consumer products and services. With vehicle OEMs, they are co-creators and implementers of codes and standards for charging and communication standards
Charging network providers	Own and operate infrastructure that provides charging capability, making them key investors in charging infrastructure. They develop new business models that match consumer preferences and provide value-added services (e.g., smart charging, bidirectional charging)
Retailers, site hosts, building owners, and charging station operators	Provide a location for vehicle charging and options for providing grid services as investors in infrastructure. They account for charging in site design and construction. They manage and optimize energy across all site loads
Standards development organizations	Convene multiple stakeholders in consensus-driven processes to develop codes and standards

(continued)

Table 2 (continued)

Stakeholder	Role
Labor and workforce education and training organizations	Develop and offer industry-recognized training and certification that provides workers with necessary technical skills, career pathways, and access to upward mobility opportunities that support recruitment, diversity, and retention
Trade associations	Support industry collaborations and coalition-building, and provide services and tools that meet member needs, convene members to develop solutions to common challenges, and represent member positions
Customers	Include residential consumers, commercial businesses, and fleets, and make the ultimate decision of the value of charging options, products, and services through their participation, which will hinge on their perceived value and convenience. They fund electricity infrastructure, and their preferences will drive new options

Another stakeholder identified in Table 2 is standard development organizations. For EVs, a key standard development organization is the Institute of Electrical and Electronics Engineers (IEEE). The IEEE is currently working on standards that begin to address the issues and provide answers to many questions including:

- How is vehicle-to-grid (V2G) technology controlled by the grid?
- How is permission given for vehicles to charge or discharge at any given time?
- How will aggregated EV batteries operate as a single power source?
- What communication protocols will be used [4]?

The IEEE eventually foresees fast charging plazas that are user-friendly—as reliable and available as gas stations currently are. This will require the integration of a wide variety of digital technologies interfacing with each other and the power grid on the other side of the meter. Operational coordination is also required. Sophisticated energy management systems will be a must as will interoperability. These systems are under development and will facilitate the ease of usage and functionality of EVs [4].

The Alliance for Automotive Innovation (AAI) advocates for the development of standards and policies that would expand the usage of existing vehicle telematics—where telematics refers to the integration of telecommunications and informatics for computing—for remote charge management that would facilitate VGI and vehicle data for billing purposes. The AAI reports that most new light-duty vehicles and commercial fleets already have many sensors and microprocessors. In addition, for more than 25 years now, telematics systems have enabled in-vehicle navigation, remote diagnostics, and roadside assistance [5].

Several utilities have undertaken pilot projects or have programs in place to test the use of vehicle telematics. These include Xcel Energy, Duke Energy, Pacific Gas & Electric, National Grid, Consolidated Edison, and DTE [5].

Examples of the type of work to be undertaken by utility regulators are the periodic reports prepared by the California Energy Commission on assessing electric vehicle charging infrastructure [6].

5 Does It Matter When EVs Are Charged?

I first studied electric vehicles when I was a wet-behind-the-ears engineer at Duke Power Company in the 1970s. One of the first assumptions that we made in our analysis at that time was that the EVs would be charged at night. Why is that?

Electric utilities are obligated to provide electricity on demand by whomever demands it, whenever it is demanded. This is called the legal obligation to serve. It means that the utility must build whatever infrastructure is required to serve that load—across all three categories of utility assets—generation, transmission, and distribution. If EV charging increases the peak demand, utilities must find additional generation resources (and associated transmission and distribution assets) to meet that demand. Over the one hundred plus years that we have had electricity, this has resulted in what is called the load curve (Fig. 1)—the aggregation of the demand for electricity from every load—industrial, commercial, residential, streetlights, whomever.

This load curve shown in Fig. 1 demonstrates the shape of the curve during the summer months—the peak demand generally occurs around 5–6 p.m. The shape of the winter load curve is different reflecting the differences in the time of day when heating is required (nights and early morning) and cooking loads around noon. Utilities can be summer-peaking or winter-peaking or their peaks can be relatively equal in both seasons. Note that the load curve never goes to zero which reflects the reality that electricity is used 24 h a day—there are uses in the home (clocks, televisions, refrigerators, coffee makers), for street lights, and for 24-h a day businesses and organizations (think hospitals, airports, police, fire, and grocery stores). The low level could be at 30% or so of the peak demand. For purposes of this chapter, I am going to assume a summer-peaking utility.

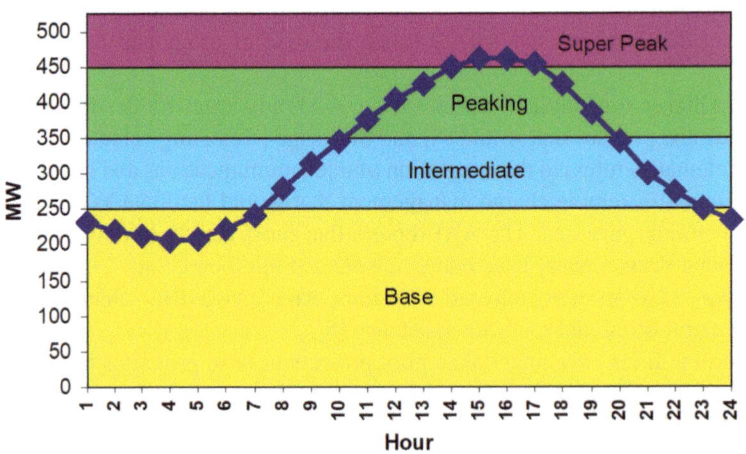

Fig. 1 Typical utility load curve. (From the author's personal files)

Studies undertaken by the California Energy Commission (CEC) of the number of chargers needed to support EVs in 2030 and 2035 in just California forecast expected power demand by hours associated with those chargers and EVs (1 million chargers for 7.1 million plug-in EVs in 2030; 2.1 million chargers for 15.2 million plug-in EVs in 2035). Figure 2 is the projection of statewide power demand for light-duty vehicle charging on a typical weekday in the year 2030. Peak demand for charging is about 4000 MW which occurs between 11 a.m. and 1 p.m. Although utilities are expected to implement time-of-use (TOU) rates, which may move charging load away from utility peak demand hours, the TOU rates are not expected to completely mitigate coincidence of charging loads with the hours of utility peak demand. The step increases in load in the evening correspond to the implementation of TOU rates causing customers to time the onset of charging. It is expected that steps will need to be taken to manage the resulting spikes. Solar energy is expected to offset much of the midday charging requirements [6].

My dream of smart chargers for EVs synced with the utility's demand curve has not come to pass. Although some locations within the USA are putting into place rate structures that incentivize customers to charge their EVs at certain times of the day or night like California's envisioned TOU rates, Brazil is taking a more proactive stance. The National Electric Mobility Platform (PNME) estimates that Brazil will need 150,000 charging stations by 2035 for a fleet of 3 million EVs and plug-in

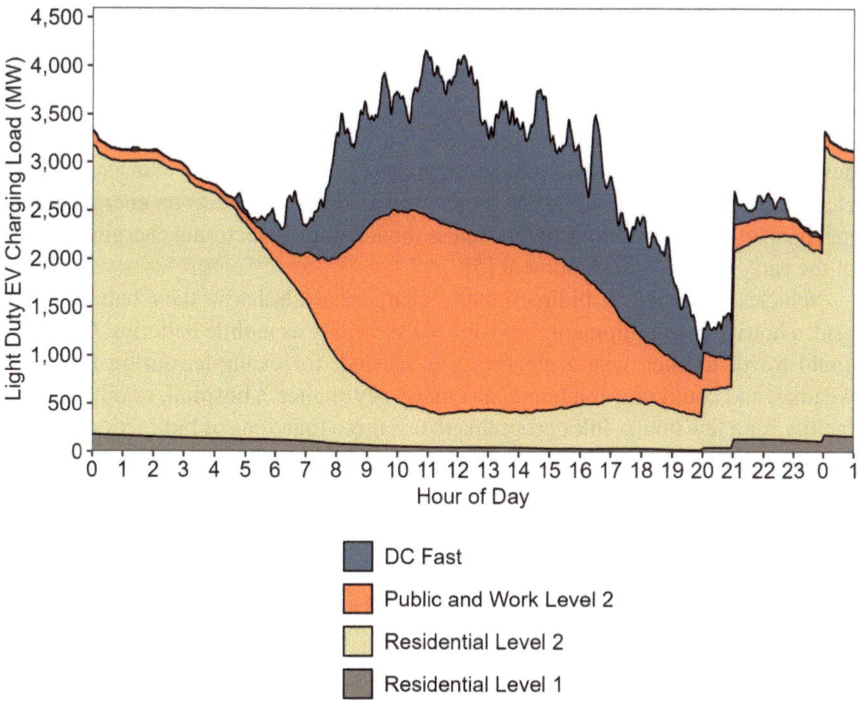

Fig. 2 2030 Projected electric vehicle charging profiles in California [6]

hybrid vehicles. Because this leads to the potential for high demand due to simultaneous charging, Brazil is implementing an intelligent charging system. The intelligent system communicates with the chargers, estimates instantaneous power consumption, and controls the maximum energy level delivered to charging vehicles simultaneously [7].

6 Can EVs Serve as a Storage Mechanism?

Hundreds, thousands, or hundreds of thousands of EVs in the same area with the capability to supply power to homes or to the grid are at a scale (MW) that becomes of interest to many of the stakeholders identified earlier. EVs become energy storage, potentially mobile energy storage, which can provide benefits to the grid including stabilization and maybe even frequency support, voltage support, congestion management, and power flow support. But in order to do this, some challenges need to be addressed [4].

One challenge is proprietary technologies. Although steps are being taken in the right direction, unlike the standardized outlets in our homes (although some are two-pronged and some are three-pronged), all EVs cannot charge at all charging stations because the technology isn't compatible among manufacturers. There is currently no universal charger and no standard charger plug-in/port hardware. And most hardware is not capable of bidirectional charging which will be necessary for V2G technology [4]. Other challenges include appropriate business models, policies, and technical standards [5].

The AAI advocates for vehicle telematics as well as aggregators to accelerate the development of bidirectional vehicle programs (meaning not only charging of vehicles from the utility but the usage by the utility of the vehicle as an energy storage mechanism). The AAI notes that business models for bidirectional charging are still in the early stages of development [5].

Vehicles equipped to bidirectionally charge can discharge their battery to the grid, a house, or to equipment. Envisioned essentially as mobile batteries, these EVs could travel to sites where electricity is needed, for example, during inclement weather, and could power a home, an emergency shelter, a hospital, or other critical facility for a few hours. Pilot programs to test these functions of bidirectional charging are ongoing at Pacific Gas & Electric, Consolidated Edison and Duke Energy, among others [5].

7 The Impacts of EVs on the Distribution System

There is never a free lunch, and EVs are no different. Can the existing distribution system handle EVs as is or will some or significant modifications be required? As with so many areas of life, the answer is "it depends." Based on the experience of

Palo Alto, California to date, factors that influence the answer include the rate of modernization of grid components, the number of EVs sold, how fast those purchases occur, where the EV owners reside, what kind of EV charging is available and used, regulatory decisions, political decisions, and economics [8].

The electric system in Palo Alto is a municipal system with a distribution system designed to meet the electric loads of the 1950s and 1960s when heating, water heating, and cooling were primarily natural gas based. Because it was designed mid-twentieth century, it was not designed for bidirectional power—just power provided by the City to customers. An associated issue is the age and capabilities of the 3150 existing distribution transformers [8].

The utility's staff currently estimates that if the City hits its target for EV production and electrical appliance saturation by 2030, 95% of its residential transformers will be overloaded. Here is the math: A 37.5 kVA transformer typically serves 15 households with a system designed for each house to draw 2 kW of power. Installing a new heat pump to replace a gas furnace draws 4–6 kW. A Level 2 charger for an EV draws 12–14 kW. Without smart meters providing insight into household peak loads to the utility, a cluster of changes could overload or blow out a transformer especially if that transformer is near the end of its life—and most of the transformers in the City are [8].

The distribution system itself needs a complete upgrade. This requires money, staffing, and technology. Until the upgrades can be accomplished, the utility must deal with confused and angry customers who are being encouraged by the City to buy EVs, charging ports, and electrical appliances but who can't currently be accommodated [8].

The situation experienced in Palo Alto is not unique. Almost 800 cities across the USA are facing similar challenges. Research from Georgia Tech found that multiple L2 chargers on one distribution transformer could significantly reduce its life—by up to 90%. Some of the life reduction results because transformers are passively cooled devices—designed to be cooled at night. If, instead, EV charging occurs at night, the transformer runs hot, reducing its expected life span [8].

What can be done? Better transformers and smart metering systems are possible answers. And a real consideration is that all solutions take time and money—and in many cases—regulatory changes utility-by-utility, state-by-state [8].

8 How Much Time Does the Grid Have to Get Ready?

Since 2017, automakers and their battery partners have committed more than $125 billion to EVs in the USA. In mid-2024, there were 113 EV models for sale at U.S. dealerships—expected to exceed 150 models within a few years. EVs exist for a wide range of automotive categories—every class of light-duty vehicle including pickup trucks, SUVs, and minivans; medium and heavy-duty trucks, buses, and off-road vehicles. About 10% of sales in 2024 were EVs although sales have plateaued [5].

How many EVs currently are in operation in the USA and around the world? How rapidly is their usage expected to increase? The International Energy Agency (IEA) reports nearly one in five cars sold worldwide in 2023 were electric with 95% of those sales in China, Europe, and the USA. These new car sales amount to almost 14 million new EVs, and the IEA reports a total of 40 million EVs currently in service [9]. With a total of 1.475 billion total vehicles worldwide in 2024, this means that 2.7% of the total vehicle fleet consists of EVs [10].

Studies have been and continue to be conducted to predict the level of EV sales in future years and how rapid that growth will be. One such study for the IEA forecasts that under certain conditions, EVs could represent 60% of vehicles sold in 2030 requiring a tremendous number of additional chargers [11]. We know that the European Union and certain states in the USA have passed legislation mandating sales of zero emissions cars only by a certain date [12]. The true answer is that no one knows and the recent reticence of Americans to purchase EVs, but a willingness to purchase hybrids, could be a harbinger of things to come.

The AAI sees EVs as "the enabler of reliable and affordable reliability" with significantly increasing electrification of the transportation sector in the twenty-first century. However, many actions must be taken to make this future a reality. These include modifications to utility rates and rate structures, increased use of data analytics, and methods of recognizing the benefits of V2G power all of which are reflected in the stakeholders and roles presented in Table 2 [5].

9 What Other Mechanisms Are Envisioned to Charge EVs?

EV charging from an external source that happens while vehicles—light-duty and heavy-duty—are moving can occur through one of three methods [12]:

- Catenary system
- Conductive (ground-based) system
- Inductive system

The catenary system uses overhead wires to provide electricity. A special kind of bus or train connects to the wires. Thus only heavy-duty vehicles can use a catenary system [12]. Catenary systems are used around the world including for the Oslo, Norway tram system; the Northeast Corridor for Amtrak (USA); Budapest, Hungary metro; and a number of systems in Japan.

Conductive charging can be used by both heavy-duty vehicles and cars using a conduction system such as a rail. The vehicles are charged through a stick that touches the rail [12]. These systems include subway systems throughout Europe, Asia, Africa, North America, South America, and Australia.

Inductive charging uses special equipment buried underneath the surface of the roads and on or in vehicles and is exemplified by Electreon's Electric Road System (ERS). The inductive coils under the road are connected to a power source. When electricity passes through the coils, they generate a magnetic field. The receiving coils installed in the vehicle capture the magnetic field and convert it to electrical energy which is used to charge the vehicle's battery. The result is dynamic charging as the vehicle is moving across the road, and the charging process does not require it to be stationary [13].

The ERS fulfills the dream of just driving EVs over roads that keep the vehicles charged instead of having to plug them in overnight or at a charging station along the road during a long-distance trip. It is a similar technology to that used for wireless charging for mobile phones. The first installations of wireless electric roads are now happening or are in process in the USA, the European Union, the Middle East, and China. These projects will test the technical and economic viability of the technology [14].

The first wireless electric road in the USA was installed in Detroit, Michigan and completed in late 2023. The quarter-mile section uses Electreon's ERS. The project came in at a cost of $2 million per mile [15]. The road is safe for drivers, pedestrians, and wildlife as coils activate only if a vehicle with an approved receiver passes over it [16]. The Detroit project is one of eighteen projects currently being undertaken by [14].

Purdue University is also planning to test wireless charging technology on a quarter-mile road test bed. Construction was progressing in West Lafayette, Indiana and expected to be completed in late 2024. The Purdue technology is aimed at charging heavy-duty vehicles (trucks) that comprise a large portion of interstate traffic. Like the technology installed in Detroit, the Purdue technology utilizes transmitter coils installed underneath the normal concrete pavement (instead of asphalt pavement). The power would be sent to receiver coils attached to the underside of the vehicle [17].

It may take 20–30 years before EVs can receive the full power they need while driving at highway speeds. In addition, EV manufacturers will decide if receiver coils are incorporated into the design of the EVs they produce [17].

10 Conclusion

Integrating EVs into the electric grid brings opportunities and challenges. A myriad of stakeholders are involved, and each has a part to play as VGI becomes a reality in the next few decades. Technology for chargers, the EVs themselves, and alternative charging mechanisms are all evolving. The existing distribution system is expected to require substantial investment of talent, resources, and dollars to enable charging at home and then at a number of other possible locations. Time-of-use rates may influence the hours at which owners charge—hopefully in a fashion that does not increase the peak demand of the host utilities—which could result in the need for more generation resources. VGI will happen—the pace and cost are still to be determined.

References

1. Biedenharn, R., Hartzog, K., & Phoenix, F. (2024, October). How Cleco power is making Louisiana an EV-friendly state. *T&D World* (pp. 20–24).
2. North American Electric Reliability Corporation. (2024, January). *Potential bulk power system impact of electric vehicle chargers: Preliminary findings and recommendations.*
3. U.S. Department of Energy, EVGrid Assist. (2024, July). *The future of vehicle grid integration: Harnessing the flexibility of EV charging*. DOE/EE-2820.
4. Wolf, G. (2024, September). Interoperability makes V2G more flexible. *T&D World* (pp. 12–14).
5. Alliance for Automotive Innovation. (2024). *Vehicle grid integration: The convergence of the automotive and electric power industries*. Retrieved November 6, 2024, from https://www.autosinnovate.org/posts/papers-reports/VGI%20White%20Paper_2024.pdf
6. California Energy Commission, Staff Report. (2023, August). *Assembly Bill 2127 electric vehicle charging infrastructure assessment: Assessing charging needs to support zero-emission vehicles in 2030 and 2035*. CEC-600-2023-048.
7. Bianchin, C. G., Schmal, R. M., Miggiorin, R. M., da Costa, R. F., de Medeiros, L., & dos Santos, R. B. (2024, March). Brazil kick-starts electric highway initiative. *T&D World* (pp. 44–48).
8. Charette, R. N. (2022, November 28). Can power grids cope with millions of EVs? *IEEE Spectrum.*
9. International Energy Agency. *Trends in electric cars*. Retrieved November 7, 2024, from https://www.iea.org/reports/global-ev-outlook-2024/trends-in-electric-cars
10. Hedges & Company. *How many cars are there in the world in 2024?* Retrieved November 7, 2024, from https://hedgescompany.com/blog/2021/06/how-many-cars-are-there-in-the-world/
11. International Energy Agency. (2022). *By 2030 EVs represent more than 60% of vehicles sold globally, and require an adequate surge in chargers installed in buildings*. Retrieved November 6, 2024, from https://www.iea.org/reports/by-2030-evs-represent-more-than-60-of-vehicles-sold-globally-and-require-an-adequate-surge-in-chargers-installed-in-buildings
12. Min, R., & Euro News. (2023). Sweden is building the world's first permanent electrified road for EVs to charge while driving. Retrieved November 6, 2024, from https://euronews.com/next/2023/05/09/sweden-is-building-the-worlds-first-permanent-electrified-road-for-evs-to-charge-while-dri
13. Vishnu, M. (2024). *Norway takes electric vehicle charging to the next level: The world's first wireless charging road.* Retrieved November 6, 2024, from https://medium.com/@bishnunath365/norway-takes-electric-vehicle-charging-to-the-next-level-the-worlds-first-wireless-charging-road-e70ad843fab5
14. Electreon. *Building the infrastructure to support mass EV adoption*. Retrieved November 6, 2024, from https://electreon.com/projects
15. BBC. (2024). *Wireless charging: The roads where electric vehicles never need to plug in.* Retrieved November 6, 2024, from https://www.bbc.com/future/article/20240130-wireless-charging-the-roads-where-electric-vehicles-never-need-to-plug-in
16. Arminas, D., & ITS International. (2023). *Electreon completes first US dynamic charging project.* Retrieved November 6, 2024, from https://itsinternational.com/its4/its9/news/electreon-completes-first-us-dynamic-charging-project
17. Albert, K. (2024). *Building the first highway segment in the U.S. that can charge electric vehicles big and small as they drive*. Purdue University. Retrieved November 6, 2024, from https://www.purdue.edu/newsroom/2024/Q2/building-the-first-highway-segment-in-the-u-s-that-can-charge-electric-vehicles-big-and-small-as-they-drive/

Jill S. Tietjen, PE, entered the University of Virginia in the Fall of 1972 (the third year that women were admitted as undergraduates after a suit was filed in court by women seeking admission) intending to be a mathematics major. But midway through her first semester, she found engineering and made all of the arrangements necessary to transfer. In 1976, she graduated with a BS in Applied Mathematics (minor in Electrical Engineering) (Tau Beta Pi, Virginia Alpha) and went to work in the electric utility industry.

Galvanized by the fact that no one, not even her PhD engineer father, had encouraged her to pursue an engineering education and that only after her graduation did she discover that her degree was not ABET-accredited, she joined the Society of Women Engineers (SWE) and for more than 45 years has worked to encourage young women to pursue science, technology, engineering, and mathematics (STEM) careers. In 1982, she became licensed as a professional engineer in Colorado.

Tietjen started working jigsaw puzzles at age 2 and has always loved to solve problems. She derives tremendous satisfaction seeing the result of her work—the electricity product that is so reliable that most Americans just take its provision for granted. Flying at night and seeing the lights below, she knows that she had a hand in this infrastructure miracle. As an expert witness, she worked to plan, license, and put into rate base new power plants.

Her efforts to nominate women for awards began in SWE and have progressed to her acknowledgment as one of the top nominators of women in the country. Her nominees have received the National Medal of Technology and the Kate Gleason Medal; they have been inducted into the National Women's Hall of Fame and state Halls including Colorado, Maryland and Delaware and have received university and professional society recognition. Tietjen believes that it is imperative to nominate women for awards—for the role modeling and knowledge of women's accomplishments that it provides for the youth of our country.

Tietjen received her MBA from the University of North Carolina at Charlotte. She has been the recipient of many awards including the Distinguished Service Award from SWE (of which she has been named a Fellow and is a National Past President), the Distinguished Alumna Award from the University of Virginia, the Distinguished Alumna Award from the University of North Carolina at Charlotte, and the Distinguished Engineering Alumni Award (special category as a non-alum) from the University of Colorado at Boulder. She has been inducted into the Colorado Women's Hall of Fame and the Colorado Authors' Hall of Fame and elected to the National Academy of Construction. Tietjen sits on the board of Georgia Transmission Corporation and served for eleven years on the board of Merrick & Company.

Her publications include the bestselling and award-winning book *Her Story: A Timeline of the Women Who Changed America* for which she received the Daughters of the American Revolution History Award Medal and *Hollywood: Her Story, An Illustrated History of Women and the Movies* which has received numerous awards. Her award-winning book *Over, Under, Around, and Through: How Hall of Famers Surmount Obstacles* was released in 2022. The memoir of Dr. Antonia Novello, *Duty Calls: Lessons Learned from an Unexpected Life of Service*, which she co-wrote, was released in 2024 in both English and Spanish versions. She is the series editor of the Springer Women in Engineering and Science series, has authored two volumes in the series, served as co-author for a few volumes and as co-volume editor for four more volumes, and written chapters for a number of volumes.

Fueling Dreams, Powering Change: A Journey Across Energy and Purpose

Wini Rizkiningayu

1 Oil and Gas Days: From Building Foundation to a Reconstruction

If you had asked me at the beginning of my career where I would end up, I could not have predicted this journey. It has been a path shaped by resilience, deep introspection, and an evolving sense of purpose: from drilling sites in Indonesia to international climate convenings, anchored by a belief that at the center of our economic activities, is the need not only to survive but also for a better, dignified livelihood.

I began my career in 2007, straight out of Gadjah Mada University in Yogyakarta, Indonesia, with a degree in Electronics and Instrumentation. As a kid, I survived from one public school to another, one scholarship to another. I was a very determined kid—joining debating team, doing multiple English public speaking competition (that led me to a pretty good savings from the winning prize), organizing and leading science club, winning bronze medal in traditional martial arts. I was practically doing whatever I can to prove my merit to the system. Coming from a lower-middle-income family, where my parents worked tirelessly to provide despite financial hardship, I was raised with the value of responsibility as the eldest child. When my father, a mechanical electrical engineer with specialization on commercial building construction, lost his job during the Asian financial crisis and never bounced back since, I watched my life change as Indonesia's political scene shifted from the authoritarian new order to a reformed government in 1998. As a first year in junior high school, I witnessed how the 'reformation' movement comes with civic unrest, riots and military responses that drives international companies out of Indonesia due to safety issues, causing engineers like my father to lost their job. My lesson to survival was when my mother was taking the helm in our family, giving room for my

W. Rizkiningayu (✉)
Rocky Mountain Institute, Jakarta, Indonesia
e-mail: wini.rizkiningayu@rmi.org

K. J. Bunker (ed.), *Women in Electrical Engineering*, Women in Engineering and Science, https://doi.org/10.1007/978-3-032-00133-7_7

father to deal with unemployment. She took cooking orders, sold jamu and dessert, and went back and forth from Tanah Abang one of the biggest textile markets, to sell clothes in my neighborhood even when she was pregnant, and doing whatever it took to keep us afloat. Witnessing her responding to crisis imparts what was probably one of the most important lessons in my life: one of adaptability and grit. While I sold some of her home-made dessert at school and accompanied her back and forth to the market on the weekend, I found my father taking respite from his grief by retelling me the story on how he project managed constructions for skyscrapers around Jakarta; the buildings in the main streets that are standing tall until now. He often explained how the electrical wiring works, how he helped design the air conditioning and plumbing that enables a building to function, and many other interesting stories of sensors, systems, and human workers around it. *He must've missed this*, I thought, recounting multiple times in the past when he took me to his construction sites and showed me around. That memory of resilience, resourcefulness, and how my family cultivated my interest to STEM has never left me. One thing for sure during my junior and high school time was: *I want to be an engineer.*

Studying electronics and instruments as my major with full-ride scholarship from Deutscher Akademischer Austauschdienst (DAAD) - German Academic Exchange Service, I often envision myself becoming not only 'just' an engineer, but one with exciting path aheads, working in the most interesting places. During my undergrad, I was very active in university-level english debate scene, a continuation of my high school pastime. Through debating, I balanced my rational-logical thinking rigor of my study with the understanding that most things in this world are grey – it's up to us what shade of grey we sees it as. University debate was when I learned about issues that I would never had learned otherwise, such as the ethics of technology, millennium development goals – now sustainable development goals, and capital inequality. It thought me how to deliver a compelling argument, how to approach an issue critically, but most of all it taught me of so many interesting fields outside of science and engineering that I'd like to explore. But being a very pragmatic person, at that time I thought I would only want to pursue work in the sector that would enable me to support my younger siblings and my family, and I will dedicate my life there. When I received a job offer from Schlumberger even before I was graduated, it felt like a dream comes true: a ticket not just to an exciting career, but to lifting my family toward a better life. Oil and gas (O&G) was one of Indonesia's highest-paying sectors, and the thought of being female field engineer seemed so adventurous and exhilirating (Box 1).

Box 1

Based on data from Indonesia statistic center—Badan Pusat Statistik (BPS), the oil and gas sector in Indonesia employed approximately 47,000 workers in 2007. This figure encompasses both upstream (exploration and production) and downstream (refining and distribution) activities. While this number represents a small fraction of the total national workforce, the sector's economic significance was substantial, contributing significantly to government revenues and exports (Mining Statistics of Petroleum and Natural Gas 2003–2007).

In 2007, Indonesia's booming oil and gas sector, driven by high global oil prices and renewed exploration, created rare and lucrative opportunities for fresh graduates, especially engineers. **Service and Oil & Gas** companies like **Schlumberger**, **Halliburton**, **Baker Hughes**, and others were **very aggressively recruiting field engineers** in Indonesia and globally, with a very early initiative on balancing male with female engineers. Factors attributing to the trends are:

- Boom in oil prices: Crude oil prices were surging (reaching $100/barrel by late 2007). That created a huge demand for more exploration and drilling activities, especially in places like Indonesia, Malaysia, and Australia.
- Aging workforce: Globally, the oil and gas sector was facing a "crew change" problem—many skilled workers were retiring, but there weren't enough younger engineers yet.
- Local content requirements: The Indonesian government was beginning to push harder on companies to hire and train more Indonesian nationals for technical roles, including field engineers.
- Skillsets targeted: They focused heavily on mechanical, electrical, civil, and petroleum engineering graduates, and even some geophysics/geology graduates depending on service lines (like Wireline or Drilling & Measurements).

These companies run a very systematic recruitment from prestigious university like UI, ITB, and UGM, offering fast-track Field Engineer roles with high salaries (IDR 5–15 million/month) and international career paths. For many women entering a male-dominated industry, these roles offered not just financial independence but a chance to break barriers, gain hands-on leadership experience in the field, and shape a more inclusive future for Indonesia's energy sector.

For 3 years, I worked as a Measuring and Logging While Drilling (M/LWD) Engineer at Schlumberger, stepping into a world far larger and tougher than the university labs I had known and spent my days in. From the Mahakam Delta's offshore rigs to the sands of Abu Dhabi training center, I experienced firsthand how raw and real the energy world could be: it takes such a complex system to deliver energy and I am a cog in the big machine. My work in drilling operation had played an essential role in setting up a foundation of work ethics; the rigor, the attention to details and the ability to adapt as I go. As female in a male-dominated industry, I learned that I have to work twice, three times as hard as my male counterparts just to gain the recognition. I learned how to mingle and still keeping boundaries. I learned that I had to adapt my 'language' to signal strength, capability and gaining acceptance. Sometimes it got toxic, sometimes I had to push my emotion away, sometimes it means putting a stone cold facade and swallowed my pride when a tool pusher was screaming at me for being slow during borehole assembly make up while what I wanted to do was making sure that everything was in order. Often the only woman at remote rig sites, I adapted to the relentless rhythm of helicopter rides, 30-day deployments with its own crew dynamics, and troubleshooting complex tools under pressure.

One night on a rig in offshore Mahakam Delta, a key sensor and cable system failed mid-operation. After more than 48 h tagging team with almost no sleep—reinstalling, rewiring, recalibrating—I stared exhaustedly at the drilling logs with my team and making sure everything was in order, knowing that a single mistake could compromise the entire operation. That night taught me a lesson no classroom ever could: behind every barrel of oil, behind every petrochemical-based product, lies an invisible tapestry of human effort (with probably sleepless nights, exhaustion and longing for home), complex precision, and grit.

Despite my background in electronics and instrumentation, stepping into the drilling world demanded a new language: one that I learned painstakingly through every rig-up, every recalibration, every report delivered under scrutiny. Imposter syndrome was a constant companion in those early years, but so was a growing competence tha, slowly but surely earned through painstaking mistakes and accumulated operation hours.

It was during this time that Total E&P Indonesia became one of my clients. Recognizing my field expertise and operational instincts, they invited me to join their geoscience team as a Petrophysicist and well log analyst. At first, I hesitated. I have zero geoscience background and my understanding of petrophysics was so limited to the formation evaluation logs as my deliverables. But the team leader at that time convinced me that they have the whole team with geology background - what they were lacking was someone who knows their ways around logging tools and its intricacies and this was where I came in. As long as I am curious enough to learn and adapt, I should be fine, he said. At Total, I entered a new chapter of active, experiential learning, but also a deep imposter syndrome. Imagine someone who had zero knowledge or even aware of what sequence stratigraphy means, joining a team of expert geoscientists in one of the most productive gas field in Indonesia. From learning basic geology and petrophysics Archie equation to being able to run probabilistic models as a daily jobs, this period of my career was one that I really stretched my cognitive capacity and put hours of learning new discipline of science. From knowing almost nothing to being able to combine petrophysical model, core and image log, I somehow proved to myself how one can adapt and extend their knowledge into a new field. However, most days I still felt like I was just 'playing as' a petrophysicist and didn't feel like I belong in a team of competent petrophysicists with geoscience background. Years later, when I was offered a junior geologist role despite not holding a formal geoscience degree, it felt like a quiet affirmation: growth, after all, belongs to those willing to learn beyond what they were trained to do.

In 2014, prompted by both personal reasons and the move of my then-partner-turned-fiancee to Jakarta, I made a pivotal career transition when I joined Paradigm

Geophysical (now AspenTech). Not only that I was moving from my comfort zone in a small O&G town in Borneo to the chaotic hustle bustle of Jakarta, I was also moving into a new role as technical sales for their formation evaluation software in a new environment that later would allow different parts of myself that I had forgotten for 6.5 years to come to the forefront. It turned out that the pieces of myself I had nurtured quietly over the years ever since high school—the love of building stories and compelling argument, the thrill of public speaking, and the joy of connecting with people—were not separate from the technical skills I had honed in Schlumberger and Total. They were waiting, patiently, to be integrated. Under the mentorship of a very intentional and strategic leader, who remains a career mentor to this day, I learned to tap into my full potential: a part of myself I had not fully accessed before, having been so immersed in technical learning and survival in the early stages of my career. As a technical sales, I combined my knowledge in formation evaluation, my drilling operation experience and my love for a good and clear narrative into consultative selling. At that time, I knew that I didn't want to just sell software; I want to excel and help my clients solve their problem. After all, I have spent years in the client's shoe when I was working in Total, working with the software day in day out. As I fell deeper into the role, I started to strategize means to deliver the software package and service around the premise of solving the client's problem and not just merely template offering. I fostered strategic relationships with clients in Asia Pacific regions through both formal engagement and my involvement in the society of petrophysicists and well log analysts (SPWLA) in Southeast Asia. I started to become a bridge between what the clients' need and what the company's research and development team can provide. This way, I wasn't just selling the software and services - I was the client's trusted advisors. This newfound synergy between my soft and hard skills propelled me forward, leading to my promotion as the Asia Pacific Business Development Manager for formation evaluation.

It was a season of flourishing. I travelled across Asia Pacific, from the bustling streets of Bangkok to the innovation hubs of Melbourne, crafting partnerships, strategizing market entries, and even developing a new software bundling package that helped the company to weather a downturn in the regional market. Along the way, I began to gradually make peace with my imposter syndrome. I stopped fighting it. I allowed it to sit beside me, a quiet passenger, while I continued to move forward, learning to believe in my own resilience. I embrace the fact that I learned geoscience by trade, not by academic training, and lean into my role in the industry. I was even actively involved in the birth of Indonesia chapter of SPWLA, setting up its governance and building the community during its initial growth stage.

Yet, beneath the surface of this professional triumph, a quiet storm was gathering and culminating in 2016. After 2 years of marriage I uncovered the painful truth of my marriage's breaking apart with infidelity. It was more than a betrayal; it was a mirror reflecting the fractures we had long ignored: the absence of quality time, the avoidance of old wounds, the silence that grew where conversations should have lived and the way we shrank to fit the mold we thought was the ideal. Not only that it was destroying the very foundational structure of my life, it also forced me into a stark crossroads: should I stay the course in oil and gas, climbing steadily toward leadership, or should I step away, tend to the shattered pieces of my personal life and reconstruct it into an unknown shape?

I fought to endure, to keep the facade intact with every workshop delivered, every software contract extended, and new clients acquired. But the truth became undeniable and it eroded into my well-being and sense of purpose. The so-called ideal life in O&G sector that has brought me to the most interesting places, one with salary that put both my siblings through college and took care of my aging parents, is no longer providing me with the solid grounds to function. The foundation I was standing on was no longer mine as what I used to believe was the recipe to a happy life: good career, decent spouse and ability to build a safety net around my family, was suddenly no longer there. And so, in 2017, amid an oil price volatility that seemed almost symbolic, I chose to step away and chose myself. I was offered a golden shake hands, as the company faced turmoil and needed to restructure due to the crashing oil price – exactly what I needed to ensure adequate support for my family and myself After a decade in oil and gas where I was constantly going from one adventure to another in a very hectic cadence, I finally took a deliberate pause. I didn't have a clear map ahead of me, but I clung to a simple mantra: I may not know yet what I want, but I know what I no longer accept. Thus began my journey of rebuilding and reconstructing myself and my life, one uncertain but hopeful step at a time.

2 Transition Times: Sabbatical, Geothermal, and Academic Journal

2.1 Sabbatical: Without Exploration, We Miss Revelation

During my sabbatical, I made a quiet vow to myself: to embark on a path of self-discovery, to open every door life offered, and to rebuild my foundation from the inside out. I want to not just remember who I was, but to discover who I could become.

One of my first steps was saying yes when one of my close friend offered me to try volunteering with Hutan Itu Indonesia, a civil society organization dedicated to introducing urban communities in Jakarta to forest conservation and the green economy. I helped organize "Concert for Forest," an event where musicians, guided by their personal journeys of reconnecting with nature, shared their art and stories. It was a year of conversations—with communities, with development practitioners, and with myself, where I began to understand how market forces, conservation, and green commodities could intertwine to shape a more sustainable future. This opportunity connected me to some people who later became instrumental in my climate journey.

In parallel, I started to inquire within: what did I use to enjoy and derived joy from. English debate and public speaking came to mind. I thought to myself, I just wanted to remember how it feels to feel joy while doing something, so I searched for a Toastmaster club nearby. I often heard about it from my colleague back in the O&G days, but I never had the time. Making my way to their meeting and somehow won the best table topic, speech, I found myself later deeply involved with Metropolitan Toastmasters Club, and helping to grow one of Jakarta's largest English public speaking clubs at the time. Through my service, I rediscovered the joy of leadership, community building, and strategic communication. I found that life did not have to revolve solely around work and the idea of ideal family. That community, too, was a vital source of meaning. Winning several district-level public speaking competitions, I pushed my boundaries further by participating in the Junior Chamber International (JCI) competition, ultimately winning the Asia Pacific title in Mongolia and becoming a finalist in Amsterdam world championship in 2017. Aside from the achievements, I learned about the power of narrative and story, and how our world is indeed built by them.

Curiosity became my first guide. I explored wedding planning, helping a dear friend orchestrate her special day; an experience that unexpectedly led to supporting other weddings over the years. I dipped my toes into the startup world, engaging with founders and helping shape a social media marketing consultancy for one of Indonesia's largest tourism state-owned enterprises. Despite all of the shiny distractions, something inside me that used to be filled with the illusion of happiness now is an inescapable gaping void. One that no matter how much interesting, new activities I've done, somehow still felt so empty and harrowing.

Amid all these seemingly random adventures, I often felt disoriented – like I was not here nor there. And somehow, it prompted me to learn about a certain state;vone where you're no longer in the old reality but not yet fully in the new one. Then came the realization: I was in a liminal space, the in-between. The anthropologist Victor Turner used it to describe rite of passage, moments when people are suspended between one social identity to another. Liminal space is unstable and ambiguous, it create openings that normal life rarely does. My sabbatical stripped me away from my usual routines, roles, and reference points, which feel uncomfortable but also turned out to be liberating. Not only that this moment had been been offering me with freedom of exploration and experimentation, but it also offered me a necessary pause to look inwards, to unlearn and relearn my pattern, and reclaim my choice-making power. A revelation: yes, the only way out is through, but how you get through matters a lot. I came to understand that this liminal space is my catalyst to a deep transformation, only if I navigated consciously. To allow for the emergence of a new, hopefully more aligned self, I need to take: the time to truly look inwards: using the opportunity to be with the gaping void that stored my pain and suffering and not just numb myself with outwards actions. Through the year, I made a commitment to really get to know myself: who I am truly, why am I here, what moves me, and many existential questions that laters became my other guides through silent meditation retreats, therapy sessions, and intentional reflection, I realized that one of my greatest fears during the crucible of my personal life wasn't just losing a relationship. It was the painful recognition that it wasn't just my marriage that had crumbled; it was my sense of self. I was afraid of being alone because

I did not know whether I even liked myself. I had forgotten how to love myself. I had lost touch with what moved me, what inspired me, and how my work and outward actions could once again become an authentic extension of who I was inside.

Through this journey of reconnection, a new clarity emerged – not only about myself but also about my sense of purpose. I wanted my work to not only contribute to my personal benefit and the idea of success that the society imposed on me. I want my energy, focus and time that I spent in working to be truly beneficial and impactful; if not for the betterment of the world around me, at least for people who faced the same barriers I did growing up. I wanted to work in a sector that not only leveraged my technical and leadership skills but also contributed to a better society, improving livelihoods and supporting sustainable futures. I wanted to work where my skills could drive societal well-being, not just economic profit. I was racking my brain, researching various options with its own risks and opportunities; and this was when the seemingly random activities during the sabbatical somehow made sense. I had a better visibility of what was possible for me, from sustainability to public speaking. However, I didn't want my past experience to go to waste. Then it hits me: if I wanted to couple my past expertise with my future aspiration, I needed to draw a through line between my past career and the fields I had been in and where I wanter to be. The answer, it turned out, had been within me all along: the energy transition.

Toward the end of my sabbatical, a friend I met during volunteering mentioned an opportunity with New Zealand government through their Trade and Enterprise agency, NZTE. It was a temporary opening to support their geothermal and renewable energy portfolio in Indonesia. This would be such an amazing entry to the energy transition world. Regardless, I was doubtful as the imposter syndrome kicked in. I had no idea about geothermal industry aside from the eerily similar drilling process with O&G. But if there was something that my 10 years careers in O&G thought me, it was that I can always adapt, learn and leap. Didn’t I also learn a thing or two about packaging narrative and deliver a compelling storytelling during my sabbatical? I gathered my courage to apply, since this opportunity felt like more than coincidence. Sometimes, the universe opens doors quietly, but I also got to go past through my own demons and approached the door. I chose to step through this one, carrying with me a new sense of purpose and a new sense of who I was becoming.

2.2 *Between Renewable Energy, Coffee, and Public Policy*

New Zealand Trade and Enterprise (NZTE) marked my first professional experience within the energy transition and public sector, although I knew that I came with a strong corporate and private sector lens. It turned out that it was a space where my skills naturally thrived. What was supposed to be a short contract turned into an almost-4-year chapter. It was a testament not just to the work I delivered, but to the

Box 2

Geothermal energy development presents a natural and efficient transition pathway for oil and gas professionals due to the high degree of technical overlap between the two sectors. Both industries require expertise in subsurface geology, drilling engineering, reservoir management, and fluid production technologies. Research indicates that approximately **70–80% of technical competencies** in oil and gas are transferable to geothermal applications, particularly in exploration, well completion, and production optimization [1, 2]. In the context of Indonesia—home to an estimated **40% of the world's geothermal resources** [3]—this skill alignment is especially critical. It enables the preservation of valuable technical capabilities while supporting national goals for energy diversification and decarbonization. Facilitating the movement of oil and gas workers into geothermal roles not only reduces the risk of stranded human capital but also strengthens the social and economic foundations of a just energy transition.

deep alignment I felt with the mission to help fostering renewable energy in Indonesia through geothermal development (Box 2).

Unlike my experience in Total where I learned geoscience from scratch, the learning curve in geothermal sector when I was working for NZTE was not steep. The subsurface exploration, technical rigor, and exploration system in geothermal felt familiar, thanks to my tenure in drilling, geoscience, and business development in the oil and gas sector. What was new—and transformative—was how I began to see renewable energy through a different lens: not only as a technical solution but as a driver of trade, development, and public good. In NZTE, I work with government, private sectors and associations involved in Indonesia geothermal sector. My work exposed me to the nuances of stakeholder engagement in a way private sector roles rarely demanded. Building and orchestrating relationships between ministries, embassy, private sectors, and associations required strategic diplomacy and a systems thinking approach. I learned quickly that policies and initiatives are not abstract—they are deeply human endeavors, dependent on trust, collaboration, and a shared vision for progress.

I was fortunate to work alongside people who were deeply passionate about growing the geothermal sector. It amazed me to learn that in New Zealand, geothermal energy was not only a domestic energy solution but also an exportable "commodity" through skills, innovation, and services. It reframed for me how renewable energy could underpin economic growth beyond national borders.

One of the most profound lessons I carried from this time was how indigenous Maori stewardship shaped New Zealand's approach to energy and resource management. I was exposed to the different dynamic of renewable energy development, from learning about Waitangi Tribunal process to observing active Maori participation in energy governance, I came to deeply appreciate how honoring indigenous wisdom and leadership could strengthen, not hinder, sustainability efforts. Intentional trips to New Zealand further deepened my understanding. I saw firsthand how the country nurtured a body of knowledge and technical leadership in geothermal development—building academic and professional ecosystems that not only served national interests but attracted global practitioners and students alike.

During this period, I became actively involved with Women in Geothermal (WING), earning a place in the Global Future Leaders Cohort for 2020–2021. I joined the executive committee for Indonesia's largest geothermal convention and exhibition, and remained deeply engaged with the Indonesian Geothermal Association. This was the period when I decided to start putting myself out there in a more public engagement, one that would help me understand the importance of female role models in renewable energy and how we need to build a tribe to support one another.

My time at NZTE was not isolated professionally; it was also a period of profound personal evolution. Beyond professional milestones, this was also a period of creative exploration. In 2018, I co-founded Sedjenak, a coffee shop in Central Jakarta, alongside four female co-founders and silent partners. Though the café eventually closed in 2023, it was a living laboratory for entrepreneurship, community building, and bringing dreams to life. Who would have thought that I would one day had owned a coffee shop? The creative endeavor has allowed me to look into the human consideration of the system, as well as challenges faced by small business to survive.

In 2019, I was introduced to the founder of an emerging initiative aimed at building public policy capacity for Indonesian youth: an initiative that would later grow into ThinkPolicy Indonesia. I joined as a mentor for their very first BootCamp, co-teaching a class on game theory and coal transition. What started as a single engagement soon grew into a profound calling.

Through ThinkPolicy, I found a new way to channel my experiences: mentoring young public servants, facilitating sessions on problem identification and root cause analysis—a skill I honed as an engineer—and experimenting with storytelling facilitation. Through COVID-19 times, teaching the online sessions are what gave me hope and motivation that not all is lost, and that learning democratization matters. Over the years, my involvement deepened; today, I continue to serve as an active mentor, facilitator, and advisor, with a personal passion for accelerating evidence-based and empathetic policymaking, especially in the realms of climate and energy transition.

Meanwhile, back at NZTE, my portfolio expanded. I took on the technology sector, honing my skills in design thinking, helping New Zealand start-ups finding their unique value proposition, learning about product development, and digital economy facilitation. I helped New Zealand tech companies enter the Indonesian market and gained firsthand exposure to the vibrancy of startup ecosystems. My earlier sabbatical experiences in public speaking and strategic communication came full circle, now serving a crucial role in pitching, facilitating, and building market-entry strategies.

This phase of my life nudged me steadily toward a bigger truth: through energy transition work, youth empowerment, entrepreneurship, and mindfulness practices, I realized I didn't just want to build a career. I wanted to be part of a good change:

one where I can help fostering a society where decent livelihoods, sustainable practices, and human creativity could thrive together. A realization dawned on me that we can only live and thrive by sustaining each other. Throughout my darkest time I was sustained by so many different kindnesses through friends and community, and I owe it to them to help build a thriving, kind society.

It became increasingly clear that public policy was the invisible thread running through every obstacle and opportunity I encountered. The barriers to energy transition and building a thriving society weren't just technical or financial—they were often systemic, rooted in policy gaps, governance challenges, and institutional inertia. And I knew, if I truly wanted to make a difference, I needed to equip myself to work at that level.

In 2019, two mentors entered my life who would help crystallize this next step. The first was Dr. Yanuar Nugroho, a former University of Manchester lecturer who had returned to Indonesia to dedicate himself to public service, serving as Deputy Chief of Staff to the President in the Executive Office of the President. The second was Ahmad Yuniarto, an Oxford scholar, then-Chairman of Schlumberger Indonesia—my first employer—and a seasoned leader in Indonesia's energy sector. Both became invaluable companions on my journey, encouraging me to see that public leadership was as much about heart and narrative as it was about technocratic structure and policy. Both encouraged me to pursue higher study in the UK with the support of Chevening scholarship.

With their support and my own growing conviction, I set my sights on further study abroad. It wasn't an easy journey. I had worked for over 13 years by then, but I wasn't sure if i was cut out for academia; I remembered back in my Oil and Gas days, I used to have a post-it next to my monitor that said "Oxbridge/Ivy League by 30 y.o.". I wasn't 30 anymore, but I knew that deep down somewhere I wanted to prove to myself that a girl from a lower income family in Bekasi can make it into those prestigious universities. Essay-writing exercises, policy literature reading, and all the preparation for scholarship was my daily grind for over a year since 2019. In 2020, I was awarded the prestigious Chevening Scholarship, a program with a competitive acceptance rate of just 2% globally. Due to COVID-19, I had to defer for a year to support NZTE operations in then-locked down Jakarta and New Zealand, running online convenings, trade missions, and being boots on the ground for various New Zealand businesses. But those lockdown months allowed me to prepare myself better and save for the next journey as I earned admission to the Master of Public Policy (MPP) program at the University of Oxford's Blavatnik School of Government. Until now, sometimes I still pinch myself that I got into one of the most selective policy programs in the world, a professional degree that would later prove to be pivotal in my career evolution.

Leaving NZTE was not just a professional transition; it marked a deeper personal evolution. I was no longer just building projects, I was beginning to build a vision of the world I wanted to help shape. And alongside this professional shift, I made an equally important personal decision: to close a chapter of my life that no longer served mutual growth.

During this time, I chose to end my marriage. After years of trying to fix what doesn't work, of self-discovery of my own strength, and of practising compassion to self, to my then-husband, and to our marriage, I finall took a decision. Not rooted not in anger, but in the recognition that both my ex-husband and I needed different environments to honor our truths and thrive. We approached our uncoupling with intentionality and respect, allowing space for both of us to move forward with kindness and clarity. This wasn't an easy decision, but a realization that love comes in so many ways and it transforms. As my career had undergone its own transformation with the Master's Degree decision, I found that my heart is also transitioning to a form that I did not think was possible before: a love and kindness for myself first.

At the heart of these transitions was a growing understanding that true systems change, whether in society or within ourselves, demands more than technical excellence; it demands courage, empathy, and the willingness to reimagine what is possible.

2.3 *Oxford Days: When a Girl from Small Outskirt Jakarta Suburb Found Herself in Bodleian Libraries Writing Papers on Climate Justice*

Stepping into Oxford was not just the next academic milestone; it was stepping into a space of profound transformation. After years of technical work and systems engagement, I was ready to examine the deeper questions: How do I want to live my life meaningfully? How can the work I do become an authentic extension of my values, my beliefs, and my evolving personal model for change? But to answer some of these questions, I had to go through the core question that had haunted me for most of my waking hours: am I good enough to be here?

The Master of Public Policy (MPP) program at the Blavatnik School of Government was an extraordinary crucible for these reflections. I was privileged to be surrounded by an inspiring cohort of peers, professors, and faculty members—all of us united by the belief that the world could be better led, better governed, and better served.

Throughout my year at Oxford, I engaged in both academic and personal discourses that sharpened my thinking not just around climate action and just transition, but about the necessity of systems thinking and intersectionality. No system, I realized, stands alone; every solution, every policy, every initiative is interconnected. Climate action, especially, is not an isolated problem—it is deeply intertwined with economic structures, social justice, and political realities. The question of privilege and inequality becomes core in my journey. I was in a reckoning of tension: I have been a case of exceptional social mobility, and I found myself in the centre of academic excellence that enables me to do so much more. How do I use this gained knowledge and social capital to enable change that is systemic, one that would enable the 10-year-old me to dream big and to know that there is a tangible way to reach her dream.

Mentors like Professor Tom Hale and my circle of "climate warriors"—classmates equally committed to driving change—and other colleagues who are passionate about driving system changes created spaces and community for me to hold and explore the tension between economic growth, social equity, and urgent climate action. I learned that land issues and climate justice are inseparable in South Africa. I learned that refugee and international border issues will be imminent in our near future with the climate crisis. I learned multiple intersectionalities of climate and energy transition that I have never been exposed to before. My time in Oxford changed me in a way that I never thought possible: the real privilege and treasure trove comes from really connecting with the brilliant friends that I would never had the chance to encounter before, understanding the world from multiple lenses and remembering our common human plight for equality, for decent livelihood, and for peace. The people that I am privileged enough to call friends, brain trust and support system: they have expanded my mind to hold multiple tensions, juxtaposed truth and they have expanded my heart through so many stories, brave conversation and reminded me again and again that being human is messy but also intricately beautiful. All of that serves as a wakeup call that solving systemic, "super-wicked" problems like climate change demands not only intellectual rigor and openness to explore various perspectives, but also demands emotional resilience, community, and sustained collective motivation.

Mirroring my experiences at NZTE, my policy interests at Oxford branched into two main streams: climate and energy transition, and the digital economy and technology sector. In 2022, my summer project took me to the World Bank, supporting a digital economy Project Management Office (PMO) policy recommendation for one of Indonesia's coordinating ministries. Simultaneously, I deepened my research work with the Oxford Martin School, under the guidance of Dr. Christian Meyer, a renowned labor economist, exploring the intersection of workforce skilling and energy transition. Little did I know, this research would point me to one of the most underlooked intersectionalities of energy transition and labor market: skilling.

That research resulted in a piece titled "Ensuring a Just Energy Transition in the Labour Market: TVET Readiness Framework and Comparative Studies." For me, it was more than an academic exercise. It was a reflection of my own career arc, moving from fossil fuels to renewable energy, and a call to ensure that millions more could navigate their own transitions with dignity and opportunity.

Through multiple conversations at Oxford that opened different doors, my project led naturally to my next step: joining Schmidt Futures as a Field Strategist Fellow. For over 6 months, I explored a critical but often overlooked space: the intersection between social well-being, workforce development, and energy transition in lower-middle-income countries. I conducted academic research, convened multi-stakeholder dialogues, and helped spark strategic discussions focused on workforce skilling as a vital pillar of equitable energy transition.

Through that work, I identified a clear "pre-market" niche: spaces where the private sector found no immediate economic incentive to invest, and where government efforts were limited or fragmented. It became evident that this was where the third sector with their mission-driven, catalytic nature, had a transformative role to play.

It was during one of these convenings that fate intervened. While reading an RMI (Rocky Mountain Institute) report on green jobs, I happened to glance at their career page—and there, staring back at me, was the opportunity for Regional Director for Southeast Asia.

It felt like more than coincidence. Here was a region I called home, a region brimming with the potential to decouple economic growth from carbon dependency—if only we could center the transition around people, justice, and equity. Here was a chance to join an organization I had admired for years: deeply technical, fiercely mission-driven, and strategic in shaping the global energy transition. One small doubtful voice persisted: I never had any experience in the third non-profit sector, and who am I, daring to apply as a regional director for a region as emerging as Southeast Asia? In the past, this kind of question would shut me down. But this time, I had years of practice proving those voices wrong, trying despite my imposter syndrome, and knowing which voices I should make peace with and which to listen to.

Sometimes, the universe speaks softly. This time, I was ready to listen.

2.4 Southeast Asia: Grounding the Transition to What Matters

After completing my fellowship at Schmidt Futures, and after years of technical, policy, and systems exploration, I felt a certain excitement to step into the arena: where ideas meet action, where ambitions meet systems, and where the future is quietly being built.

That's when the opportunity at RMI found me.

Today, I lead a $three million portfolio for RMI in Southeast Asia, working at the nexus of economic growth and climate action. In our region, the energy transition is not only an environmental imperative; it is about reimagining prosperity, dignity, and human flourishing. My team's mission is simple but profound: How do we help countries in the region foster economic growth while decoupling it from carbon emissions and greenhouse gas footprints?

In Southeast Asia, the energy transition narrative goes beyond "saving the planet." It is about building dignified, resilient livelihoods. It is about crafting an alternative growth model that centers humanity, equity, and opportunity. In my current role, our work focuses on deploying techno-economic insights to create market-based levers for change. Through this lens, I began to see Southeast Asia's transition as a delicate dance: navigating geopolitical tensions, trade flows, supply-demand shifts, and the unique local dynamics that define this diverse region of emerging economies.

The barriers to financing clean projects must be lowered; bankability for small, medium, and large projects must be strengthened; policy readiness and ecosystem maturity must be catalyzed. Building supply signals—through decarbonization of the power sector, green industrial strategies, and clean transport—requires more than innovation. It demands robust policymaking, business ecosystem readiness, private sector confidence, and an active effort to break information asymmetry between governments and markets.

Every day, my team and I engage across governments, private sectors, and civil society organizations, mindfully navigating sensitive spaces: captive coal for critical minerals, low carbon industry, fossil energy transition, carbon credits for transition finance, clean mobility adoption, and workforce capacity building for future leaders in energy transition.

In 2025, I was honored to be selected as a Yale Emerging Climate Leaders Fellow, further deepening my engagement at the intersection of climate action, energy transition, and social equity. This opportunity allows me to further break silos and drive collaboration with other fellow energy leaders across the Global South to help me better understand energy transition and its intersectionality in Southeast Asia. Through this fellowship, I am again reminded how climate actions, both on mitigation and adaptation, is all about humans in the centre. I get to be part of a very supportive group of emerging global south climate leaders and get involved in various discourses ranging from the power of storytelling, energy entrepreneurships, innovative policymaking, equitable and just transition intersecting with the economics of decarbonization, to the creative economy as driver of sustainability and stewardship of vernacular arts. This fellowship mirrors the complexity of our world and reminds me of my Oxford days, summarized by my favourite Maori saying: the most important thing in the world is the people, the people, the people.

We stand at a pivotal nexus: energy transition as not just an environmental necessity, but as an alternative, human-centered pathway for economic growth. But one thing remains true: energy transitions do not happen without people—including the conversation of green jobs and workforce transition (Box 3).

Box 3

A successful energy transition in Southeast Asia depends not only on deploying new technologies but also on managing a just and inclusive workforce transition. The region, still heavily reliant on fossil fuels for economic growth, is home to millions of workers whose livelihoods are tied to traditional energy sectors. According to the International Labour Organization [4], Southeast Asia could generate up to **14.2 million new green jobs by 2030**, but the skills gap remains a major barrier.

Without intentional strategies to reskill and redeploy oil, gas, and coal workers into clean energy sectors like renewables, energy efficiency, and sustainable infrastructure, the transition risks exacerbating inequality and social instability. Countries like Indonesia, Vietnam, and the Philippines are beginning to recognize that workforce policies must move in parallel with technological and financial shifts.

Building pathways for existing workers—through targeted reskilling, certification programs, and cross-sector mobility—ensures that the energy transition delivers broad-based prosperity rather than leaving legacy communities behind. Workforce transition, therefore, is not just a social imperative but a core enabler of Southeast Asia's sustainable economic future.

According to the International Labour Organization, the global shift to a green economy could create 24 million new jobs by 2030, but only if we center people, communities, and justice at the heart of every strategy. Learning from my personal arc in career transition to green jobs, enabling environment and visibility of transition pathways needs to be some of the consideration in policymaking for green jobs and in supporting career transition to a more sustainable sector.

From field engineer to regional director, my story is just one in the vast realm of green jobs and job transition. In its core, the career shifts for me is a long journey. It is a journey of carrying forward core values: solving problems with rigor, embracing active learning with humility, and practicing adaptive leadership with intention. The imperative for global energy transition—a vast, intricate system change—mirrors my personal transformation. For me, it is the story of returning to my core: redefining success from personal gain to collective flourishing.

It is about moving from environments that constrained my well-being to spaces where my values, strengths, and hopes could thrive. It is about moving from extracting energy from the Earth to nurturing both people and planet. And while my journey continues to evolve, one conviction remains unchanged: transition and transformation, be it in the energy space, career space, or personal space, begins when each of us chooses to align our actions with our deepest values, with a brave reckoning of our current reality and the future we are willing to reimagine, one courageous step at a time.

3 Reflection: Designing a Green Workforce Worth Believing In

When I chose to leave the fossil fuel sector, it wasn't because I lacked opportunities. Oil and gas had provided me with stability, technical growth, and a clear professional ladder; all things that mattered deeply, especially for someone from a modest background in the global south who had fought hard for every rung of that ladder.

But over time, I began to feel a dissonance between what I was building and what I believed in. I wanted to align my work with the future I hoped to see: a future where economic growth does not come at the cost of climate, equity, or dignity. Still, making that leap wasn't easy. It took a sabbatical, self-reflection, and support from others to step into a space that, while more aligned with my values, was less secure and far less structured. My decision-making has not always been idealistic, most of it was actually pragmatic. My personal forecast of the current energy trend, the workforce needs for certain soft and hard skills, compensation and how this energy transition sector can support my well-being, those were some of my considerations of moving to a greener job. It took certain privileges such as social capital, financial and non-financial resources, reckoning with gender stereotypes, familial support and efforts of social mobility. All the tangible and non-tangible resources that I gained while climbing the ladder, my 'first' privilege of having such a supporting family that demonstrated unconditional love, my circle of friends, mentors, coaches and community, the ability to build a safety net, these are just a few of many that allowed me the 'freedom' to move careers, to take a sabbatical, to make decisions and bets that others might not be able to. It takes a certain privilege and willingness to take the risk - two things that aren't easy for the young workforce in emerging economies to do - to choose green jobs or transition to greener jobs. Yet for more sustainable, low carbon development to happen, we need these brilliant minds to be in green jobs.

And that's the paradox we must confront. If we want more people, especially from legacy industries, to join the green workforce, we cannot expect them to make that leap on belief alone. We must design the space to be not just idealistic, but viable.

The opportunity is clear: global employment in renewable energy reached 16.2 million in 2023, surpassing fossil fuels for the first time. In the Global South, particularly in Africa and Southeast Asia, projections are just as promising: 3.3 million new green jobs could be created across Africa by 2030, with up to 70% in solar alone. The International Labour Organization estimates that a global green economy could create 24 million jobs by 2030.

But the reality behind these numbers is more complex.

Demand for green skills is growing fast, up 11.6% in the past year, but supply is lagging behind at just 5.6%. Workers from fossil industries often find few structured pathways for reskilling. And many women, especially in the Global South, face even higher barriers. Globally, only 4.6% of women in emerging economies hold green jobs, compared to 16% of men. In the G20, just 22% of STEM roles are held by women.

When I reflect on my own journey: the uncertainty of leaving oil and gas, the thrill of rediscovering purpose, and the deep learning curve of systems change; I can't help but ask: What would it take for more people like me to make that shift? What would be helpful for women, especially, in knowing that green jobs are a possible pathway for them to support themselves and their families? How can we collectively create good, green jobs that allow people to align their values to the reality of the world and at the same time the hope for building a better, sustainable future?

Reflecting from my personal journey, the barriers I faced and the enabling environment that has helped me in getting to where I am, there needs to be some shifts in the form of:

- Reskilling must become a public good: accessible, supported, and regionally tailored, especially for workers coming from carbon-intensive sectors. Upskilling and reskilling for green jobs needs to be mainstreamed and become both national and local government economic agenda, supported by good collaboration with private sectors, academics, and local community. The interventions should involve the understanding of green job creation as local economic competitiveness, the entrepreneurial opportunity and ecosystem creation for good green jobs, lowering the gap of demand and support signals in the labour market, democratizing both in-class and hands-on learning for skills needed in green jobs, giving the agency back to local communities to co-define green jobs alongside the market, and protection around workers' rights especially in career transition as a consequence of energy transition.
- Green careers must be visible, stable, and aspirational: green jobs should not be positioned as sacrifice, but as opportunity to lead and grow. The risk and uncer-

tainty around deciding to take on green jobs should be lowered through policy, incentives, protections, and overall healthy market creation to open and sustain more green jobs. When green jobs are at a level playing field with conventional economy and fossil energy jobs, choosing to transition to greener sectors will have lower barriers.
- Gender inclusion must move beyond quotas: there needs to be a focus on transforming the work culture and leadership structures that keep women from thriving in technical and policy spaces. It requires dismantling legacy systems built on technocratic hierarchies and masculine norms that equate authority with control and expertise with dominance. Institutions must shift from rigid, positional leadership to adaptive, networked models that value collaboration, systems thinking, and relationship intelligence as core competencies. This transformation depends on embedding gender equity into the structures of performance, financing, and leadership development, while redefining merit to reward inclusivity and collective problem solving. Leaders and institutions must create psychological safety and enabling environments in the form of tangible support to multiple roles that women have to assume in their life, and shared accountability so women can question norms and shape strategy without penalty. The energy transition is not only technological but deeply cultural, and its success depends on changing who leads, how power is exercised, and what values define progress.
- Young people must be engaged not just as future leaders, but as co-creators of today's solutions—given room to innovate, question, and build differently. There needs to be a shift where young people must be positioned not just as beneficiaries of the energy transition but as co-creators shaping it in real time. Institutions and leaders need to move beyond token youth engagement toward genuine power sharing, allowing young professionals to question legacy systems, experiment with new eco-systems and start up models, and design alternative pathways for green growth and just employment. This requires transforming mentorship into co-creation, integrating youth voices in decision-making, incentivizing and redefining innovation as intergenerational collaboration in real life. By giving young people the authority to build differently, the transition gains the creativity, legitimacy, and resilience needed to turn structural ambition into societal transformation.

To truly achieve a just transition, we cannot only decarbonize our economies—we must humanize our workforce systems. That means supporting people through career transitions as intentionally as we support industries through technological ones.

While it took me over a decade and a long personal detour to align my work with my values and to be in a change space where my work and drive is valued fairly, the current wold does not have that much time. The climate crisis is looming, and the imperative for human-at-the-center of sustainable development is more pressing than ever. We need workforce, brain power, and human capital to realize a thriving humanity, yet good green jobs are scarce, risky, and with high barriers of entry. Skilling and education need time to catch up with climate solutions, yet the need to build climate resilience is now.

Choosing to transition to green jobs shouldn't have to be that hard.

A green economy must offer more than clean power.

It must offer people, especially women, the marginalized community, and the next generation, the chance to thrive.

References

1. Lowry, T. S., et al. (2020). *The role of oil and gas expertise in accelerating geothermal energy development*. Sandia National Laboratories.
2. Tester, J. W., et al. (2006). *The future of geothermal energy: Impact of enhanced geothermal systems (EGS) on the United States in the 21st century*. MIT Press.
3. International Geothermal Association. (2021). *Global geothermal energy overview 2021*.
4. International Labour Organization (ILO). (2022). *Skills for a greener future in Southeast Asia*.

Wini Rizkiningayu is the Southeast Asia Regional Director at RMI (Rocky Mountain Institute), where she leads Southeast Asia portfolio focused on equitable energy transition, climate finance, and workforce development. A Chevening Scholar and Oxford MPP graduate, Wini brings over 15 years of experience spanning oil and gas, renewable energy, public policy, and system change. She is also a Yale Emerging Climate Leaders Fellow for 2025 cohort, committed to building a future where climate action drives human dignity, resilience, and collective prosperity.

Antenna and Its Applications

Amna Ikram

1 Introduction and History of Antenna

Antennas are critical components in modern communication systems, serving as the interface between guided waves, such as those in transmission lines, and free space. They convert electrical signals into electromagnetic waves, enabling wireless communication across vast distances. From the simple dipole antenna to complex phased arrays, antennas play a fundamental role in various applications including radio broadcasting, satellite communication, and cellular networks [1].

1.1 The Evolution of Antenna Technology

The history of antennas is an interesting journey that involves physics, engineering, and communication. From the earliest time experiments with electromagnetic waves to the sophisticated designs of today, antennas have played an essential role in shaping modern communication. This section explores the key developments in antenna technology, highlighting the innovations and thinkers who have contributed to its evolution. The concept of wireless communication began to take shape in the late nineteenth century. In 1888, Heinrich Hertz conducted experiments that confirmed the existence of electromagnetic waves, laying the groundwork for wireless technology. Hertz's apparatus included a simple dipole antenna, consisting of two metal rods that emitted radio waves. Although Hertz himself did not envision practical applications for his work, his findings inspired many inventors and scientists.

In the early 1900s, Guglielmo Marconi utilized Hertz's principles to develop the first practical wireless communication system. Marconi's work involved refining

A. Ikram (✉)
Mason, OH, USA

K. J. Bunker (ed.), *Women in Electrical Engineering*, Women in Engineering and Science, https://doi.org/10.1007/978-3-032-00133-7_8

antenna design for improved transmission and reception. His use of the vertical monopole antenna and the grounded, resonant loop were groundbreaking. By 1901, Marconi succeeded in transmitting a signal across the Atlantic Ocean, demonstrating the prospective of radio communication. As radio technology progressed, so did the need for more efficient antennas.

In the 1920s and 1930s, there was a significant advancement with the introduction of the Yagi-Uda antenna by Hidetsugu Yagi and Shintaro Uda in 1926 [2]. This directional antenna became a key in radio broadcasting due to its high gain and ability to focus signals on specific directions, significantly improving reception quality. The arrival of television in the 1930s further pushed antenna innovation. The dipole and Yagi antennas [3] adapted for television reception became common household items, allowing millions to access broadcasts.

World War II accelerated antenna development, motivated by military communication needs. Researchers created new designs such as the log-periodic and parabolic antennas, which offered enhanced performance for radar and other applications [1].

Post-war, the booming consumer electronics market fueled further advancements. The rise of cellular technology in the 1980s introduced a new era for antennas. Multiband antennas were developed to handle different frequency bands, accommodating the growing demand for mobile communication.

The late twentieth and early twenty-first centuries ushered in the digital revolution, profoundly affecting antenna technology. The transition to digital broadcasting necessitated the creation of antennas that could handle higher frequencies and broader bandwidths. Technologies like MIMO (Multiple Input Multiple Output) antennas emerged, allowing for multiple signal paths to enhance data transmission rates [4].

Recent developments in antenna miniaturization and integration with microelectronics [5] have paved the way for innovative applications. Today, antennas are embedded in smartphones, IoT devices, and vehicles, enabling ubiquitous connectivity.

The history of antennas reflects a continuous quest for improved communication methods. From Hertz's early experiments to today's advanced technologies, antennas have evolved dramatically, driving innovations in various fields. As we look to the future, the importance of antenna technology will only grow, facilitating the next generation of wireless communication and connectivity. The journey of the antenna is a testament to human ingenuity and the relentless pursuit of knowledge, shaping how we connect and communicate in an ever-evolving world. The evolution of antennas dates to the late nineteenth century with the pioneering work of inventors like Heinrich Hertz, who demonstrated the existence of electromagnetic waves, and Guglielmo Marconi, who successfully utilized antennas for long-range communication. Over the decades, advancements in materials, design, and theoretical understanding have transformed antennas, leading to their integration into numerous technologies.

1.2 *Importance of Antennas in Modern Communication*

In today's digital age, antennas facilitate critical services such as mobile telephony, Wi-Fi, GPS, and satellite communication. The demand for higher data rates and more reliable connections drives ongoing research and innovation in antenna design. Understanding the principles and applications of antennas is essential for engineers and researchers working in telecommunications, aerospace, and many other fields.

1.3 *Classification of Antennas*

Antennas can be classified based on various criteria, including shape, operating frequency, and application. Common types include:

1. Wire antennas which have a simple structure and are often used in short-range applications.
2. Array antennas composed of multiple elements to enhance performance.
3. Microstrip antennas that are flat and lightweight, ideal for integration with circuits.
4. Reflector antennas that are used in applications like satellite dishes to focus signals.

2 Aircraft Antenna

In aircrafts, antennas are used for communications as well as for different navigation systems. These navigation systems provide several functions for the operation of the aircraft, such as Direction Finding (DF), Distance Measuring Equipment (DME), Global Positioning System (GPS), Microwave Landing System (MLS), Radar Altimeter (RadAlt), etc. [6].

Figure 1 [6] lists different communication and navigation systems used by an aircraft along with respective antenna placement location. Large aircraft duplicate essential systems like Very High Frequency Communication (VHF Comms) system, Traffic Collision Avoidance System (TCAS), and Air Traffic Controller (ATC) system for redundancy and safety reasons. There are several challenges associated with aircraft antennas due to physical structure and placement.

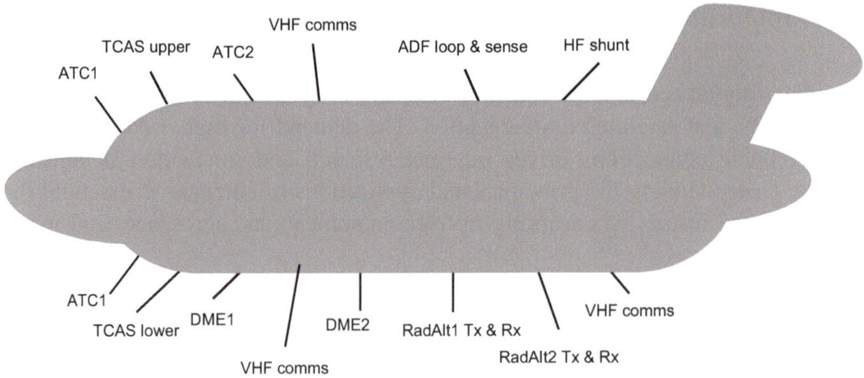

Fig. 1 Layout of aircraft antennas for various communication systems [1]

2.1 Size of Antenna at Low Frequency

Commercial aircrafts rely on blade and whip antennas which are highly visible and obtrusive as shown in Fig. 2. Some of the problems associated with currently employed aircraft antennas are large size and their maintenance. When the frequency of operation is low, the wavelength becomes large which makes the physical antenna size prohibitively large for most applications and unacceptable for platforms such as airplanes [7]. Typically, a blade antenna is a quarter-wavelength monopole antenna encapsulated in a radome. A large external antenna has a greater tendency to collide and get damage with the ground equipment especially during

Fig. 2 Blade and whip antenna protruding outside the aircraft surface

aircraft maintenance and service. In aircraft, for the aerodynamics and the mechanical stability, external antennas are a sensitive factor.

2.2 Impact of Antenna on Air Drag

During flight, airstream flows around an antenna. The physical interaction of the antenna with the air medium around it causes an aerodynamics drag and generates aerodynamics noise [8]. The improper placement of antenna, along with additional weight and size of an aircraft antenna, can increase the aerodynamics drag, which negatively affects the flying ability of an aircraft, such as increasing fuel consumption which eventually puts limitations on the speed of aircraft.

The compact and conformal antenna helps to find an installation location within the boundary layer especially if the antenna is added after the aircraft shape was frozen, i.e., normally a case where an additional installation or an upgrade to the aircraft electronics communication system is performed. A boundary layer is defined as a very thin layer of air flowing on the surface of aircraft. It gives an effective shape to any surface, which is slightly different from the actual physical shape. The details of the flow within the boundary layer are very important to understand the skin friction drag of an object. As any object protruding outside the boundary layer can produce a large drag force on the aircraft and cause turbulence [6].

2.3 Multiband Antennas

The main advantage of combining two or more functions in one antenna results in the reduction of the total number of antennas required on an aircraft. This helps to minimize the overall aerodynamic drag, reduce the number of service repairs or replacements due to corrosion and foreign object damage, and reduce the number of structural holes needed to install the antenna [7]. However, if a multifunctional antenna fails, it can cause failure of more than one service thus additional measures are required to maintain reliability.

2.4 Impact on Antenna Performance Using Composite and Aluminum Aircraft

For most of the antennas placed on the aircraft, the skin of the aircraft works as a ground surface. Although the aircraft fuselage is cylindrical, the surface is usually very large and acts like a flat ground plane for an antenna. In the past, aluminum was solely used to make the aircraft body but in last few years, composite aircraft has gained the market's attention and is a replacement to the aluminum material. The main advantage of using composite material for an aircraft is the weight reduction, improved fuel economy, and improved material strength as compared to aluminum material [9]. Composite materials are more resistant to corrosion as compared to aluminum and therefore have low maintenance and repair cost associated with them [10]. The composite material does not break easily as compared to aluminum, which can easily bend and dent after the damage. So, it is harder to find the internal structural damages in composite. Unlike aluminum, the reduced conductivity of composite structure does not make them a good candidate for electromagnetic shielding such as in-case of lightning strike [11]. Composite materials are expensive due to complex fabrication processes. Efficient planes are made up of more than 50% of lightweight carbon fiber composites [10]. In an aircraft manufactured by using composite materials, antenna grounding is performed by using an aluminum sheet under the antenna and feed [9].

3 Medical Sensor Antennas

In the modern healthcare system, medical sensors and antennas play a crucial role. They enable wireless communication and data transmission for monitoring and diagnostics of various vital health issues [12]. Some of the key medical sensors are as follows. Wearable sensors are used for continuous monitoring of vital signs such as heart rate, temperature, and glucose levels. Implantable sensors can monitor conditions internally, such as blood pressure, glucose levels, oxygen saturation, and heart rate [13]. Environmental sensors detect changes in the environment, such as air quality or temperature, affecting patient health.

Different type of antennas based on their size and radiation characteristics are used in medical devices, such as

1. Patch antennas are compact and suitable for wearable devices, providing good performance in a small form factor.
2. Dipole antennas are simple and effective, used in various medical devices for short-range communication.
3. Loop antennas are useful for RFID applications in medical tracking systems.

The important features are size and form factor as antennas need to be small and unobtrusive for wearable or implantable devices. Typically, Wi-Fi and ISM bands (2.4 GHz) are used for pairing the device. The materials used for such sensors must be safe for human interaction, especially for implantable sensors. Having lower efficiency is critical for battery-operated devices to prolong operational life.

The real-time tracking of vital signs to detect anomalies in patients is an important application of wearable antennas [14]. By providing telecommunication channels between patients and healthcare providers, they provide smart health systems. By focusing on these aspects, medical sensors and their antennas can significantly enhance patient care and health monitoring.

4 Base Station Antennas

Base station antennas are critical components in modern telecommunications systems, enabling wireless communication for mobile devices and IoT applications. They serve as the interface between the network infrastructure and the end users, providing essential connectivity and coverage as shown in Fig. 3. This section explores the types, design principles, and advancements in base station antennas, highlighting their significance in the wireless landscape.

Base station antennas can be categorized based on their design, functionality, and application.

4.1 Types of Base Station Antenna

1. Omnidirectional antennas radiate signal uniformly in all horizontal directions. They are commonly used in urban environments where coverage is needed in all directions. They are suitable for small cells, public safety networks, and rural deployments where a 360° coverage is beneficial [1].
2. The other type is directional antennas which focus energy on specific directions, providing higher gain and extended range. They are ideal for point-to-point

Fig. 3 Cluster of telecommunication tower-antennas at elevated area for maximum coverage. (Picture taken at Steptoe Butte State Park, WA)

communications, backhaul links, and areas requiring targeted coverage, such as stadiums or large venues. Another type is a sectoral antenna. This type of directional antenna divides a coverage area into different sectors such as 60° or 90° wide. They are designed to serve multiple users within a defined sector. They are frequently used in cellular networks to enhance capacity and manage traffic in densely populated areas [5].

3. MIMO (Multiple Input Multiple Output) antennas employ multiple antennas at both the transmitter and receiver to improve communication performance and data rates [4]. They play an integral role to modern LTE and 5G networks; MIMO antennas increase throughput and enhance reliability by utilizing spatial diversity.

4.2 Effective Design of Base Station Antenna Array Systems

Designing effective base station antennas involves several key principles such as gain and radiation patterns. The gain of an antenna affects its coverage area and signal strength. Understanding the desired radiation pattern is essential to ensure optimal performance in the intended environment. Another important parameter is polarization; antennas can be vertically or horizontally polarized. The choice of polarization impacts signal quality and should align with the mobile devices being used in the area. Base station antennas must be designed for specific frequency bands (e.g., 700, 1800, 2600 MHz) to comply with regulatory requirements and optimize performance.

Base station antennas must withstand various environmental conditions, including wind, rain, and temperature extremes. Materials and designs should prioritize durability and reliability. The field of base station antennas has seen significant advancements driven by the need for increased capacity and coverage. Understanding the complexities of base station antennas is essential for engineers, network planners, and stakeholders in the ever-changing landscape of wireless communications. Some of latest technologies and system deployed are as follows.

4.2.1 Active Antenna Systems (AAS)

AAS integrate multiple antennas and RF components in a single unit, allowing for advanced features like beamforming and dynamic adjustment of coverage patterns.

4.2.2 Smart Antenna Technology

Utilizing algorithms and adaptive techniques, smart antennas can dynamically adjust their radiation patterns in response to user demand and network conditions, optimizing performance in real time.

4.2.3 Integration with 5G Networks

The rollout of 5G technology demands higher frequency bands and more sophisticated antenna systems, including massive MIMO [15] and beamforming capabilities to support dense urban environments and high user capacity.

4.2.4 Sustainable Designs

As environmental concerns grow, there is an emphasis on designing antennas that minimize energy consumption and incorporate sustainable materials, contributing to greener telecommunications infrastructure.

Base station antennas are foundational to the efficacy of wireless communication systems, directly impacting coverage, capacity, and user experience [15]. As technology evolves, ongoing innovation in antenna design and functionality will continue to shape the future of telecommunications, enabling smarter, more connected environments.

5 Handheld Device Antennas

Handheld devices have become an integral part of modern life, transforming the way we communicate, work, learn, and entertain ourselves. These devices are compact, portable, and powerful, enabling users to perform a wide array of tasks on the go. From smartphones and tablets to portable gaming consoles and e-readers, handheld devices are ubiquitous in daily life. This part of the chapter explores the evolution, design, functionality, and future trends of handheld devices, providing insights into their impact on society and technology.

The mobile phone revolution in the late 1990s and early 2000s marked a significant shift in handheld devices. Along with the smartphones and tablets, there are specialized handheld devices designed for several other functions. These include portable gaming consoles like the Nintendo Switch, e-readers like the Kindle, handheld GPS devices, digital cameras, and medical devices.

The field of designing mobile handset antennas has become increasingly challenging as the requirements of highly efficient multiband coverage rise versus the limited available volume. With the evolution of 1G to 4G/LTE technology, the dual band antenna has evolved into a penta-band antenna [3]. With the advancement to 5G technology, and evolving 6G, a cellphone now has a phased array antenna [4].

In the case of internal handheld antennas, the physical lengths associated with the available volume are typically small with respect to the wavelengths of the lower cellular frequency bands. The electrical length of the antenna can be increased by folding and bending the structure in a definite order to get multimode resonant frequencies [16]. In a cellular handset, the ground plane is the main radiator at the lower frequency bands while at higher frequencies, the internal antenna pattern itself is electrically large enough to become an efficient radiator.

Planar Inverted F Antenna (PIFA) and loop antenna are widely used in mobile handheld devices. It has a self-resonating structure with purely resistive impedance at the frequency of operation. The tuning parameters to change the electrical performance of PIFA are the height of the antenna with respect to the ground plane, path length of the resonator, and the distance between the feed point and short point [17]. The patch, slot, and stub are used to compensate for the mismatch, enhance the matched bandwidth, and improve the radiation characteristics [18].

To effectively design and utilize antennas, one must understand fundamental concepts such as radiation patterns that define the directional distribution of radiated energy. The ability of antenna to focus energy on a particular direction is known as the directivity. Impedance matching ensures the efficient power transfer between the antenna and the transmitter/receiver, and the bandwidth determines the range of frequencies over which an antenna operates effectively. All the antennas in handheld devices must meet certain standards as defined by regulatory authorities, in different parameters such as Efficiency (η), Specific Absorption Rate (SAR), Hearing Aid Compatibility (HAC), and body effects.

6 Phased Array Antennas

Phased array antennas are a crucial technology in modern communications, radar, and remote sensing applications. Unlike traditional antennas, which physically rotate to change the direction of their beam, phased arrays rely on electronic control to steer their beams rapidly and accurately.

A phased array antenna consists of multiple antenna elements arranged in a specific geometric pattern. These elements are fed with signals that are phase-shifted relative to each other, allowing the antenna's radiation pattern to be electronically controlled [19]. By adjusting the relative phases of the signals, the direction of the main beam can be steered without moving the physical array.

The key idea behind a phased array antenna is constructive and destructive interference. When the signals fed to the individual elements are in phase, their radiated signals combine constructively at certain angles, forming a strong beam in that direction. When the phases of the signals are varied, the direction of constructive interference shifts, steering the beam in the desired direction [20].

6.1 Phase Shifting and Beam Steering

The ability to steer the beam electronically is achieved by adjusting the phase of the signal fed to each antenna element. If an element is fed with a signal that is ahead in phase by a certain amount, it will radiate a signal that is delayed relative to other elements. By changing the phase difference systematically across the array, the direction of the beam can be steered across a wide range of angles.

In true phased arrays, the phase of the signal is adjusted at each individual antenna element using phase shifters [19]. This provides the most flexibility and control over the beam, allowing both beamforming and beam steering in any direction.

In frequency scanning arrays, the beam direction is controlled by adjusting the frequency of the transmitted signal. The array does not use phase shifters for beam steering, but instead relies on changes in frequency to shift the beam [20].

6.2 Antenna Elements and Feed Network

The fundamental components of a phased array antenna are the individual antenna elements. These are typically simple antennas, such as dipoles, patch antennas, or slot antennas, arranged in a grid pattern. The type of element used depends on the application, as different elements have different radiation patterns, bandwidths, and efficiencies.

The feed network is responsible for distributing the signals to the individual antenna elements. It must ensure that the signals are delivered with the correct phase and amplitude to each element. The feed network can be either analog or digital, with digital feed networks being increasingly common in modern systems due to their flexibility and precision [21].

6.3 Phase Shifters

Phase shifters are the key components in a phased array system that controls the phase difference between the signals fed to the individual antenna elements. These devices can be implemented in various ways, such as using mechanical, electronic, or optical methods, depending on the frequency and application of the antenna [22].

6.4 Beamforming Network

Beamforming is the process of shaping and steering the radiation pattern of the antenna array. The beamforming network ensures that the signals fed to each antenna element are in the correct phase and amplitude, enabling the formation of a coherent beam in the desired direction [19].

6.5 Advantages and Challenges of Phased Array Antennas

Phased arrays can rapidly steer the beam electronically, without the need for mechanical movement. This allows for much faster and more precise tracking of targets or communication links. With appropriate signal processing, phased arrays can generate multiple beams simultaneously, which is particularly useful for applications like radar and satellite communication. Unlike traditional mechanically steered antennas, phased arrays have no moving parts, leading to greater reliability and lower maintenance. Phased arrays can achieve fine resolution and high accuracy in beamforming, making them ideal for radar, sonar, and communication systems where precise control over the beam is required [21].

The design and implementation of phased array systems can be complex, particularly when it comes to the integration of phase shifters, feed networks, and beamforming systems. Phased array antennas tend to be more expensive than traditional antennas, due to the advanced technology required to manufacture and implement them, including phase shifters, control electronics, and array elements. The active components in a phased array antenna, such as the phase shifters and beamforming network, can consume significant amounts of power, which may limit the size and efficiency of the system.

Phased array antennas are extensively used in radar systems for both military and civilian applications. Their ability to steer beams quickly allows for continuous tracking of fast-moving targets without mechanical scanning. This capability is critical in applications like air traffic control, weather radar, and military defense systems.

In satellite communications, phased array antennas allow for rapid beam steering, enabling the dynamic formation of communication links between satellites and ground stations. This is particularly important for low Earth orbit (LEO) satellites, which move quickly relative to the Earth.

Phased arrays are employed in electronic warfare systems for jamming, electronic surveillance, and radar spoofing. Their ability to switch beam direction rapidly makes them ideal for disrupting adversary radar and communication systems.

6.6 Future Trends in Phased Array Technology

The continued development of materials science, microelectronics, and computational techniques is likely to drive significant advances in phased array antenna technology. Future trends include miniaturization; as electronics continue to shrink in size, phased array systems will become more compact, allowing for integration into smaller platforms, such as drones and autonomous vehicles. Advances in digital signal processing will enable even more sophisticated beamforming techniques, allowing for finer control over the beam and improved performance. The future of phased arrays may feature adaptive beamforming, which adjusts the beam in real time based on environmental conditions, such as interference or changes in the target's position.

Through their ability to electronically steer beams, generate multiple beams, and operate with no moving parts, phased arrays enable significant advancements in a wide range of applications [22]. As technology continues to evolve, the role of phased array antennas will only grow, driving innovation across numerous industries.

References

1. Balanis, C. A. (1997). *Antenna theory analysis and design* (2nd ed.). John Wiley & Sons.
2. Ramsay, J. (1981). Highlights of antenna history. *IEEE Antennas and Propagation Society Newsletter, 23*(6), 7.
3. Stutzman, W. L., & Thiele, G. A. (1997). *Antenna theory and design*. John Wiley & Sons.
4. Bird, T. S. (2016). *Fundamentals of aperture antennas and arrays: From theory to design, fabrication, and testing*. John Wiley & Sons.
5. Balanis, C. A. (2012). *Advanced engineering electromagnetics* (2nd ed.). Wiley Global Education.
6. McNamara, T. M. (2010). *Introduction to antenna placement and installation*. John Wiley & Sons.
7. Hale, F. J. (1984). *Introduction to aircraft performance, selection, and design*. John Wiley & Sons.
8. Ikram, A. (2018). *Low-profile antennas for aircraft communication systems*. PhD dissertation. Ecole Polytechnique.
9. Balanis, C., & DeCarlo, D. (1982). Monopole antenna patterns on finite size composite ground planes. *IEEE Transactions on Antennas and Propagation, 30*(4), 764.
10. Birken, J. A., Wallenberg, R. F., & Milton, O. (1981). Advanced composite aircraft electromagnetic design and synthesis. In *IEEE International Symposium on Electromagnetic Compatibility, Boulder, Colorado, USA*.
11. Blake, C. L. (2016). Composites—Their electrical and electromagnetic impact. In *International Symposium on Electromagnetic Compatibility, Washington, DC, USA*.
12. Bhatia, D., Paul, S., Acharjee, T., & Ramachairy, S. S. (2024). Biosensors and their widespread impact on human health. *Sensors International, 5*, 100257.
13. Vo, D.-K., & Trinh, K. T. L. (2024). Advances in wearable biosensors for healthcare: Current trends, applications, and future perspectives. *Biosensors, 14*, 560.
14. Rahmat-Samii, Y., Topsakal, E., & Moghaddam, M. (2021). Antenna and sensor technologies in modern medical applications. *IEEE Antennas and Propagation Magazine, 63*(5), 148.
15. Chen, Z. N., & Luk, K.-M. (2009). *Antennas for base stations in wireless communications* (1st ed.). McGraw-Hill.
16. Ikram, A., Beckman, C., & Irmscher, S. Design and development of a multiband loop antenna for cellular mobile handsets. In *International Workshop on Antenna Technology (iWAT), Hong Kong, China, March 2011.*
17. James, J. R., Hall, P. S., & Wood, C. (1981). *Microstrip antenna theory and design*. The Institution of Engineering and Technology.
18. Geyi, W., et al. (2008). Handset antenna design: Practice and theory. *Progress in Electromagnetics Research, 80*, 123.
19. Mailloux, R. (2017). *Phased array antenna handbook* (3rd ed.). Artech.
20. Yang, G., Zhang, Y., & Zhang, S. (2021). Wide-band and wide-angle scanning phased array antenna for mobile communication system. *IEEE Open Journal of Antennas and Propagation, 2*, 203–212.

21. Visser, H. J. (2005). *Array and phased array antenna basics*. Wiley.
22. Kim, J., & Lee, H. L. (2022). High gain planar segmented antenna for mmWave phased array applications. *IEEE Transactions on Antennas and Propagation, 70*(7), 5918–5922.

Amna Ikram, PhD, received BS from University of Engineering and Technology, Pakistan, and MS from Högskolan i Gävle, Sweden, and the PhD from Ecole Polytechnique de Montreal, Canada (2018), all in electrical engineering. She has worked in Laird Technologies, Kista and GE Global Research Center, New York as a skilled and accomplished Senior RF Engineer with expertise in antenna and RF circuits design and simulation for sub-6 GHz and mm Wave applications. She has extensive experience in leading R&D projects and developing production-ready solutions for defense, aerospace, and telecommunications industries. With a strong foundation in RF engineering, leadership, and innovation, she excels in bridging research with real-world RF solutions for next-generation defense, telecom, and aerospace systems.

Breaking Barriers: A 17-Year Journey in the Utility Industry

Sheena Simpson

In a world of endless possibilities, the ultimate potential often lies beyond the realms of our imagination. Seventeen years ago, I embarked on a journey that would shape me into the person I am today. From a timid, sheltered young girl in southern Belize to a leader in a male-dominated field, my story is one of resilience, growth, and a relentless pursuit of excellence.

1 Early Influences: Finding My Path

Growing up, gender was not a lens through which I initially viewed the world. The innocence of childhood allowed me to see endless possibilities without the constraints of stereotypes. Yet, as I grew older, I became increasingly aware of how societal norms influenced expectations for boys and girls. For girls, there were constant reminders of how to sit, speak, and behave "properly," shaping a framework that often limits aspirations. These societal influences often create silent boundaries, making certain aspirations seem unattainable.

My childhood in southern Belize was filled with numerous moments of observation and self-reflection. I noticed early on how adults treated boys and girls differently, often encouraging boys to be assertive and adventurous while advising girls to be nurturing and restrained. This disparity was evident in the toys we were given and the activities we were encouraged to pursue. Boys played sports and explored freely, while girls were guided toward roles emphasizing caregiving and household responsibilities. These early lessons shaped societal expectations, but they also

S. Simpson (✉)
Belize City, Belize
e-mail: sheena.kuylen-simpson@bel.com.bz

K. J. Bunker (ed.), *Women in Electrical Engineering*, Women in Engineering and Science, https://doi.org/10.1007/978-3-032-00133-7_9

planted a seed of curiosity within me—a curiosity about why things were the way they were and how I could forge a different path.

High school was where my fascination with engineering truly began. Faced with the choice of selecting an elective between agriculture and electricity, I chose the latter, drawn by its complexity and potential for innovation. Electricity became a gateway to the sciences, and I further pursued Physics and Math with fervor. Though I was often the only one among my peers who ventured into these subjects, I was undeterred. My journey was guided by logic, passion, and an unyielding belief in my potential. Each class, project, and experiment deepened my understanding and love for the sciences, reinforcing my determination to pursue a career in engineering.

My determination led me to the University of the West Indies, where I became the first Belizean female in my Computer & Electrical Engineering Bachelor's program. This was a pivotal chapter in my life, filled with challenges and opportunities for growth. Navigating a new country and academic environment alone tested my resilience. The rigor of the curriculum, coupled with cultural adjustments, forced me to adapt quickly. While my introverted nature made it difficult to share struggles, I found solace in late-night study sessions and the support of newfound friends who understood the demands of our program. Remaining focused and having the support of my family helped me to overcome those exhausting moments. Every so often I needed a pep talk from my mom to gear me back up again to get back into the battle of reaching my goal. By the end of my studies, I graduated with honors, a testament to perseverance and the guiding mantra: "This too shall pass."

2 Navigating the Utility Industry

At the age of 22, I joined the electric utility company in Belize as a young graduate eager to carve out a path in a challenging field. My roles spanned across substation operations, distribution operations, system planning, and engineering. Each role offered a unique perspective on the intricate systems that power communities, but it also exposed the gender disparities deeply ingrained in the industry. The technical challenges were only part of the equation; navigating the social landscape was equally demanding.

As a woman in a male-dominated field, I faced the challenge of constantly proving my competence. Meetings often required me to assert my ideas with a balance of confidence and tact. I recall instances where my suggestions were overlooked, only to be validated later when echoed by male colleagues. These moments were disheartening, but they also fueled my determination to make my voice heard. I learned the importance of claiming credit for my contributions and developed strategies to navigate these dynamics without compromising my authenticity. Over time, I discovered that persistence, preparation, and poise were invaluable tools for navigating this complex environment.

Leadership presented its own set of challenges. Balancing assertiveness with empathy, I embraced a management style rooted in emotional intelligence. The evolving landscape of the utility industry—marked by renewable energy integration, climate resilience, and technological advancements—demanded innovative solutions. As a Manager, I have overseen critical operations that ensured the stability and sustainability of our energy grid. This role has deepened my passion for engineering and reaffirmed my commitment to lifelong learning. The more I learned about the intricacies of the grid and its role in community development, the more I understood the broader impact of my work (Fig. 1).

2.1 *Balancing Family and Career*

While my professional journey has been demanding, my roles as a wife and mother have been equally fulfilling. My husband's unwavering support has been a cornerstone of my success, providing stability and encouragement during challenging times. Together, we have nurtured our children, creating an environment where they feel empowered to explore their unique talents and defy traditional gender norms.

The COVID-19 pandemic brought unprecedented challenges, blending work, school, and home life into a single space. I was enrolled in my master's program

Fig. 1 A visit to Utility Scale Solar Farm

online during the pandemic from 2019 to 2021. It was a difficult time for me to create a work life balance routine. With both my children and I navigating online education, the boundaries between professional and personal responsibilities blurred. These experiences highlighted the importance of adaptability and prioritization. By creating designated spaces for work and family, I managed to maintain a semblance of balance, though not without moments of feeling overwhelmed. Faith and family became my anchors, reminding me that challenges are temporary, while growth is constant.

Parenthood has also been a journey of learning and growth. Watching my children develop their unique personalities has been a source of inspiration. My daughter, who is athletic and a natural leader, defies traditional gender roles with confidence, while my son's introverted yet creative nature reminds me of the importance of nurturing individuality. Together, they have taught me to value diverse perspectives and to champion inclusivity, both at home and in the workplace. These lessons have not only shaped my parenting approach but have also informed my leadership style at work, fostering an inclusive and supportive environment for my team.

2.2 Lessons Learned and the Road Ahead

Reflecting on the past 17 years, I am struck by the transformative power of resilience. Each obstacle—from overcoming stereotypes to balancing competing priorities—has been a catalyst for growth. Programs like Women in Renewable Energy (WIRE) have provided invaluable opportunities for mentorship and skill development. Through various initiatives, I have had the privilege of guiding aspiring engineers while gaining fresh perspectives from their journeys. These exchanges have reinforced the idea that growth is a two-way street, where mentors learn as much as mentees.

Life is a series of chapters/seasons that you go through. Not everyone will get to experience all the seasons in your life, so do not expect everyone to understand your growth, passion, and determination. Each season marks different lessons learnt and challenges. As you go through the different stages in life, your appreciation and value of certain things will change. This is an expected outcome as your experiences will give you deeper insights. Each chapter will come with the victories to celebrate and the challenges to overcome. In difficult times, remember that failure builds character. It is only when you overcome failure that you're able to unlock more of your potential. There is a lot of untapped potential in yourself, so constantly push yourself beyond your defined limit.

You will not always make the right decisions, and these moments will serve as important learning steps in your career. Give yourself grace during these times. What matters most is that you approach each situation with careful thought, ensuring you have fundamentally considered the main factors and parties involved in your decision-making process.

Experience is the best teacher, and not every employer will give you the respect and recognition you deserve. When this happens, remember that options are always available. Do not be afraid of change. Explore your options and ensure you are in a place where you feel valued and can contribute meaningfully. These shifts can be transformative, enabling you to align your talents and goals with an organization that appreciates your worth.

The utility industry is at crossroads, grappling with global challenges such as climate change and the integration of renewable energy. As a leader, I am committed to driving innovation and fostering an inclusive environment that empowers the next generation of engineers. By embracing emerging technologies and prioritizing sustainability, we can shape a future that meets the needs of our communities while safeguarding the planet. My vision includes building stronger networks between countries and sectors, ensuring that resources and ideas are shared to achieve a more resilient energy future (Fig. 2).

3 Overcoming Stereotypes and Building Confidence

The journey of unlearning societal conditioning has been pivotal in shaping my identity as an engineer and leader. Cultural norms often teach girls to be nurturing and passive, while boys are encouraged to be assertive and bold. These expectations manifest in the workplace, where women's leadership styles are often scrutinized. As a young engineer, I learned to navigate these dynamics with resilience. Commanding attention in meetings, articulating ideas with clarity, and reclaiming credit for my contributions became essential skills. This required constant self-reflection and a willingness to adapt, ensuring that I stayed true to my values while navigating external expectations.

Fig. 2 Workplace Setting

One of the most striking challenges was finding the balance between assertiveness and approachability. While a confident demeanor was often misinterpreted as aggression, a softer approach could be perceived as a lack of competence. Over time, I developed a leadership style that combined strength with empathy, recognizing that effective communication and collaboration are the cornerstones of success. This balance has allowed me to lead with authenticity and inspire others to bring their best selves to the table.

4 Embracing Leadership and Mentorship

Leadership is not merely about authority; it is about inspiring others and fostering growth. I emphasize strategic planning and team development. By championing diversity and inclusion, I aim to create a workplace where every voice is valued. Mentorship has been a significant aspect of this journey, offering a platform to share experiences and support the next generation of engineers.

I had never been fond of public speaking and being in the spotlight. Naturally, I preferred to be behind the scenes, but often I am expected to represent the company at events that require both being in the spotlight and public speaking. The more prepared I am, the easier it is for me to execute these tasks. Knowing what makes you uncomfortable and finding ways to increase your level of comfort can significantly build confidence. I have worked on this aspect over time, to the point where it has become easier to execute such responsibilities. Being in sync and aligned with the position you represent on a wider stage also gives you an upper hand in confidently defending that position. It is equally important to focus on developing areas that need improvement, ensuring continuous growth in both professional and personal realms.

Through WIRE, I have had the opportunity to mentor women across island nations, exchanging knowledge and building networks that transcend borders. This experience has reinforced the importance of community and the transformative power of shared learning. As both a mentor and mentee, I have grown immensely, gaining insights that have enriched my personal and professional life. By fostering these relationships, I have witnessed the ripple effect of mentorship, where empowered individuals go on to empower others.

5 The Intersection of Family and Career

Balancing a demanding career with family responsibilities is a constant juggling act. My children have been a source of inspiration, reminding me of the importance of breaking stereotypes and creating opportunities for growth. Their individuality—from my daughter's boldness to my son's introspection—has taught me to embrace diversity in all its forms. These lessons extend beyond my personal life, influencing how I approach team dynamics and problem-solving at work.

The pandemic underscored the need for flexibility and resilience, as our home became a hub for work, school, and family life. These challenges deepened my appreciation for the support of my husband and the strength of our family unit. Together, we navigated this period with determination, finding joy in the small moments and growth in the struggles.

Parenting has also shaped my approach to leadership. Much like at home, creating an environment at work where individuals feel valued and empowered fosters collaboration and innovation. My experiences as a mother have taught me the importance of empathy, patience, and adaptability—skills that are just as vital in the workplace. By integrating these values into my professional life, I strive to lead with purpose and integrity.

6 A Vision for the Future

As the utility industry evolves, so do the challenges and opportunities. The integration of smart technologies, renewable energy, and climate-resilient infrastructure is reshaping the landscape. I am committed to staying at the forefront of these advancements, leveraging my expertise to drive meaningful change. My goal is to continue breaking barriers, fostering innovation, and inspiring others to reach their full potential.

One of the most exciting aspects of the future is the growing emphasis on diversity and inclusion in STEM fields. By creating pathways for underrepresented groups, we can ensure that the next generation of engineers reflects the diversity of the communities they serve. Mentorship programs, community outreach, and advocacy are critical to achieving this vision. Additionally, I envision a future where cross-sector collaborations drive innovation, enabling us to address complex challenges with collective wisdom and effort.

7 Final Thoughts

Life is a journey of continuous growth. The obstacles we face are not roadblocks but steppingstones to greater heights. My journey as a female electrical engineer in the utility industry has been one of discovery, resilience, and empowerment. To anyone embarking on a similar path, I encourage you to embrace your uniqueness, push past societal expectations, and remain steadfast in your pursuit of excellence. Your journey matters, and the world is waiting for your contributions.

Each chapter of my journey has been a testament to the power of determination and the value of community. By sharing my story, I hope to inspire others to recognize their potential and to approach challenges with courage and confidence.

Together, we can build a future that celebrates diversity, fosters innovation, and uplifts every voice. This journey has shown me that no dream is too ambitious, and no challenge is insurmountable with the right mindset and support system.

Sheena Simpson is a dedicated wife, mother, and seasoned professional who strives to find harmony among these pivotal roles. With over 17 years of experience in the electric utility industry, she has navigated the challenges of a male-dominated field with resilience and determination. Recognizing the importance of representation and mentorship, Sheena decided to share her personal journey—one of growth, perseverance, and self-discovery—to empower and guide other women pursuing careers in engineering. Her chapter, *Breaking Barriers: A 17-Year Journey in the Utility Industry*, chronicles her experiences as a female electrical engineer, shedding light on the obstacles she has overcome and the lessons that have shaped her into the well-rounded individual she is today. Through her story, she hopes to inspire young female engineers by offering real-life insights into the profession and encouraging them to embrace their unique paths with confidence.

Beyond her professional endeavors, Sheena finds joy in baking, playing volleyball, traveling, and cherishing quality time with her family.

Index

K. J. Bunker (ed.), *Women in Electrical Engineering*, Women in Engineering and Science, https://doi.org/10.1007/978-3-032-00133-7

MIX
Papier aus verantwortungsvollen Quellen
Paper from responsible sources
FSC® C105338

If you have any concerns about our products,
you can contact us on
ProductSafety@springernature.com

In case Publisher is established outside the EU,
the EU authorized representative is:
Springer Nature Customer Service Center GmbH
Europaplatz 3, 69115 Heidelberg, Germany

Printed by Libri Plureos GmbH
in Hamburg, Germany